Human Measurement

Human Measurement

Anthony Harris

Ph.D., F.R.I.C., M.I.Biol.

Heinemann Educational Books

To my students, past, present, and future.

Heinemann Educational Books Ltd
LONDON EDINBURGH MELBOURNE AUCKLAND TORONTO
HONG KONG SINGAPORE KUALA LUMPUR NEW DELHI
NAIROBI JOHANNESBURG LUSAKA IBADAN
KINGSTON

British Library Cataloguing in Publication Data

Harris, Anthony B.
  Human Measurement.
  1. Human growth   2. Anthropometry
  I. Title
  612.6   QP84

ISBN 0-435-60360-4

© Anthony Harris 1978
First published 1978

Published by Heinemann Educational Books Ltd
48 Charles Street, London W1X 8AH
Printed in Great Britain by
BAS Printers Limited, Over Wallop, Hampshire

Contents

		Page
	Preface	vi
	Acknowledgements	vi
	Introduction	vii
PART I	**THE GROWTH, SIZE AND SHAPE OF THE BODY**	**1**
Chapter 1	Stature and weight	3
Chapter 2	Some aspects of human growth	12
Chapter 3	Analysis and classification of physique by external form	17
Chapter 4	Sexual dimorphism	36
PART II	**FUNCTION. SOME METHODS AND MEASUREMENTS IN PHYSIOLOGY, PSYCHOLOGY AND NUTRITION**	**47**
Chapter 5	Weight, surface area and obesity	49
Chapter 6	Physiological performance	60
Chapter 7	Homeostatic mechanisms	67
Chapter 8	Perception	72
Chapter 9	Some surveys and measurements in nutrition	77
PART III	**TREATMENT OF RESULTS**	**87**
Chapter 10	Statistical methods	89
Chapter 11	Correlations in classifications of physique	97
	Appendix	105
	Index	107

Preface

I wrote this book for people interested in human biology, a subject in which important original work can still be done with the simplest of apparatus—scales, measuring tapes, the camera—providing the method of enquiry is properly planned. Here then, is an unparalleled text of highly refined scientific measuring and method using as subject material the people around you.

A unique method of classifying human physique is described, progressing from simple observation to quantitative criteria. This text can be used at any level of enquiry, as an absorbing V and VI form study, or as a tool for graduate and postgraduate work in biology, medicine, nutrition, sociology, and education.

Anthony Harris
Islington, London 1978

Acknowledgements

My first thanks must go to the thousands of people who served as guinea pigs, volunteers, and experimenters. Special thanks too for our models, and to Chris Bray, Jeanie Rosefield, Jane Wilkinson, Sue Vincent, Jane Gardner, Hilary Cooper, and Deana Trappitt for enthusiastic help in measuring and calculating and surveying.

Over the ten-year period spent preparing this work I have badgered my colleagues for facts, opinions, and apparatus; to them, all over the world, thank you. (What I have done with this help is my responsibility.)

Calculations done were lightened by the loan by Messrs Bondain of the most advanced computors as they came on the market, culminating in a Rockwell 920 card programmed calculator.

Acknowledgements to reproduce the photographs on pages 38 and 39 are due as follows: Young Soldier and St John the Baptist, by Rodin; Bachante and Cupids, by Marin: Victoria and Albert Museum.

Cupid complaining to Venus, by Cranach; Nude, by Renoir; Bacchus, by Titian: reproduced by courtesy of the Trustees, The National Gallery, London.

Introduction

Much important work in biology and medicine is not statistical. The circulation of the blood was demonstrated on one individual, and then repeated on others. Although analysis of large samples of populations are important, preliminary work can begin with an individual, or with small groups of people.

This aspect of experimental work is due for emphasis today, since with the introduction of statistical methods there has been less attention to well planned (e.g., keeping variables to a minimum) experimentation, with the hope that statistics will rescue the results. This cannot happen. We would do well to recall that Newton used mathematical analysis rather than the statistics of chance. As remarked in *New Scientist*, when it was discovered that a gross majority of papers in the *British Medical Journal* treated their results in a statistically erroneous manner:

'The nub of the matter is that well-designed experiments produce either meaningful results or negligible ones. Statistics in such cases serve only to render objective the convictions of the researchers concerned. In very few instances will statistical manipulation extract valid extra information over and above what is at once evident.'[1]

Biology, perhaps human biology especially, still needs much qualitative observation. Konrad Lorenz,[2] like Darwin, has used this simple approach to uncover important features of behaviour and adaptation in the animal world. His philosophy, although non-statistical, is still important in the progress and study of biology. The emphasis in this text is for the teacher and student to use their powers of observation, *and* to utilize numerical analysis.

Quantitative description

A science becomes of age when measurements are used. Chemistry and physics are quantitative sciences, but centuries of qualitative work preceded their numerical description of nature. In the text will be found both qualitative and quantitative methods of approach to describing the human body.

In the section dealing with classification of human physiques the interplay of simple visual inspection, classifying by form, and the final use of numbers, is seen very clearly; while the advantages of one approach against another are examined. Mathematical descriptions, although quantitative, are always theoretically possible in any branch of study, but it often appears that the exactness obtained obscures simple ideas, or indeed is so time-consuming as to make the descriptions impractical.

Statistical treatment

Measurements in biological work must be treated statistically because the values seldom are fixed. For example, whereas the number of legs we have is two, the height of our species is time dependent, sex dependent, genetically dependent, and nutritionally dependent. There is no fixed value for the height of *homo sapiens*; there is a range of heights. Providing we fix our variables of sex, age, and so forth, we can determine the values of height for a group of people, add the values up and divide by the number of people, so giving the *average* value for the group. This notion of average is used throughout the text as a simple guide.

I have kept statistical calculations out of the main body of the text because the principles of human measurement can be shown by graphs, histograms, and various other display techniques. The proper way to treat data is first to *examine* it, and *then* to apply statistical tests. Accordingly, the student who wishes to know how to transform his raw data into acceptable quantitative standards of mean, standard deviation, significance, and so on, will find an introduction to the methods in Part III where much of the data shown simply (but none the less accurately) in the text is couched in statistical terms.

Units

The metric system of measurement is used in the text. New work should always use the metric system, but since much of the literature will remain in pounds/feet/inches some values here are given in these old units. This will help the student appreciate some of the older literature, provide a firm basis for comparing the two systems, and aid in some survey work when, for example, raw data is still couched in feet, inches and pounds (e.g., the police in UK and USA still describe people in feet, inches and pounds). Furthermore, in comparison work where ratios of lengths are of interest the absolute scale is unimportant providing it is used consistently.

Measurement

At first sight, nothing is simpler than measuring a length. A ruler is put against it, and the measurement is read off. This is a basic measuring technique, and, with weighing, is at the basis of nearly any measurement you like to mention. The rules you use are based on a standard which is supposed to be constant, but the Imperial Standard Yard contracted at the rate of one part in a million every twenty-three years. Standard reproductions of the metre can be made accurate to one part in 100 000 000, but the constancy still has an uncertainty, and the length is only known to an approximation, no matter how small the uncertainty is. The ultimate standard of length is not liable to contraction and bending as a piece of platinum iridium is: the metre is now defined as 1 650 763.73 wavelengths emitted in a vacuum by krypton atoms of mass 86, under specified conditions.

These conditions include temperature, which has to be standardized, while of course the limit of a vacuum can never be absolute. The British copy of the kilogram was 1.000 000 058 kg in 1922, 1.000 000 071 kg in 1948, and back to 1.000 000 059 in 1961.

It should be clear therefore, that the notion of absoluteness in measurement is imprecise. *Any statement of quantity is always an estimate*. Usually it is possible to pick instruments and use methods which give the level of accuracy you require, but it is wise to remember that the finest Vernier scale, when examined under a microscope, fades into a chaos of pits, paint, and erosion. But even if your instruments are good enough, there are still other kinds of errors to be avoided. For example, take the measurement of height. Suppose you set out to measure teenager John's height. Let's see what can happen.

First measurement: carried out with a tape of linen. You give John a height of 175 cm, but can't decide where to put your thumb; say 180 cm.
Second measurement: next day, a modern height machine (stadiometer) complete with automatic gauge gives height measurements. Height 176.4 cm.
Third measurement: this time you note John is stooping, so you wonder 'Should I standardize my technique?'. You find that if you make him stand to attention he is taller; if you lift his head, from the front, stretching his neck a bit, you can get him to inch up a bit more. Height 176.9 cm.
Fourth measurement: three years later, same technique, 180.5 cm.
Fifth measurement: next day by your assistant, 180.1cm.
Sixth measurement: by you in the morning, 180.5 cm, in evening 180.0 cm.

Being careful, you work on the problem until you standardize everything about the whole procedure and do it exactly the same way each time and make sure that other people can do it the same way too.

The variation in John's height arose because:
(a) you did not standardize your method;
(b) other people carried out the measurement their way, not in a standard way.

Even so, true variation occurred because what you were measuring was varying. John was a teenager; you discovered he grew.

True variation can occur because of periodic changes in the subject measured. In the morning after sleep we are all of us taller than the night before. You discovered a circadian fact. But even when true variation is not present your measurements will still vary, if you keep refining your techniques.

Your results should not reflect a physically impossible or meaningless level of accuracy. Thus if you say a muscle weighs 1 kg, I imagine you to mean anything between 0.5 and 1.5 kg, but if you say it weighs 1.1 kg, then I take you to mean between 1.05 and 1.15 kg. If you say 1.01 kg, then 1.005 and 1.015 kg are the limits. On the other hand, if you say it weighs 1.0000001 kg, I do not really know what you mean, unless it is desiccated, wrapped up, and so is not gaining or losing weight. Similar remarks go for any kind of measurement. If you state a figure, then it is customary to assume you are stating a value to the best of your experimental accuracy. To say a person's height is 170 cm is not the same statement (although no doubt no one will notice the difference in eyeball–eyeball confrontation) as saying his height is 170.0 cm. Viz:

    170 cm—height between 169.5 cm and 170.5 cm
    170.0 cm—height between 170.05 cm and 169.95 cm.

Of course, there is much personal variation of etiquette here. Some researchers will tell you when they have rounded up. Some will tell you when they have not; others will give the last figure they got from their measurement, others will give so many significant figures. It is not always clearly stated, but it is wise for you to be sure of what your own method is, even though it varies from set of results to set of results. You may ask at this point, 'What has this got to do with statistics?' The answer is 'everything', because you can only apply the powerful methods of statistical analysis to data, and unless the data are defined clearly, there is no way you will get them clearer with any mathematical technique. Take a look at any ruler measured out in centimetres. Using this, measure the distance between any two convenient lines. You will have to estimate the number of millimetres, or the decimal point. You can be sure, can't you, whether the fraction is less, or more, than half a centimetre? Your results on a series of such line pairs would be of the form x cm, to the nearest whole centimetre. With a ruler with mm divisions, you can apply the principle to millimetres, your results will now be $x.y$, to the nearest whole millimetre.

If you have good eyesight, you may say you can give the tenths of the millimetre, in which case your results are given in the form $x.yz$. With a Vernier, z can be estimated easily, and you might like to estimate the thousandths of a centimetre, in which case your results would be in the form $x.yzv$. The rule, then, is measure to your limit where convenient, and express your results significantly to an order $\frac{1}{10}$th less in accuracy and decide yourself whether the last figure 5 is to be left, or rounded down, or up, e.g., reading 1.115 = 1.11 (rounded down), = 1.12 (rounded up), or left as 1.115. If the best reading you can get is 1.117, express your result as 1.117 and, say the last figure is an estimate, or express it as 1.12 (correct second significant figure). In general, 1, 2, 3, 4 are taken to add nothing to the preceding figure; 6, 7, 8, 9 are taken to add one unit to the preceding figure.

Percentage accuracy in measurements

Most people can read off a length to the nearest millimetre, while an estimate of a tenth of a millimetre is possible too, using an ordinary rule. The thickness of the marking lines on the rule are approximately 0.1 mm, so, clearly, measurements more fine than this with such an instrument are meaningless.

If a length of 1 cm is measured, since we can be sure that the millimetre estimate is correct we are justified in giving the result as 1.0 cm, while other observers may read the value as 1.01 or 0.99 cm. In other words, since the limit of our accuracy is 0.01 cm, we can measure 1 cm to $\pm 1\%$, 0.5 cm to 2%, while 0.1 cm can only be measured with a certainty of 10%, one part in ten. If we measure 100 cm, our accuracy, or uncertainty, is one part in 10 000. By using instruments to measure values much larger than their finest division we easily improve the accuracy of our results. In weighing, for example, even ordinary chemical balances are graduated to 1 mg. We can therefore weigh a gram to one part in 1000, or 100 g to one part in 100 000. If a final result involves the multiplication or division of two quantities, the uncertainties of which are $x\%$ and $y\%$ respectively, then the final result is uncertain by at least $(x+y)\%$.

In biological work the statistical variation is usually so much larger than these technical errors, that measurement errors are usually not computed; nevertheless, since the mean value depends on individual results, these should be obtained with as much accuracy as convenient.

Double blind experimentation

Darwin was so aware that we tend to select what fits our preconceptions, he made special attempts to note facts which seemed to contradict what he was trying to prove. This tendency to bias is so strong you need to be careful how you conduct your experiments. The general term for the method used to eliminate this type of bias is the double blind experimentation.

Although the technique varies according to the study undertaken, it really involves separating: selection of subjects to be measured, measuring the subjects, and evaluating the results. In practise this would usually mean a large team of experimenters, and so fully double blind trials are rarely achieved. For example, if you want to correlate, say, height with ability to do the long jump, and you feel tall girls jump better than short ones, you may be unconsciously tempted, when picking your study group, to choose girls who jump well and who are tall, and girls who jump badly and are short. Your results will, of course, 'prove' your theory. To avoid this error the subjects are not told why you are measuring height or long jump performances. Indeed the person who measures height should be a different person to the one who measures the jump. The two sets of results are later paired by another individual, preferably again by someone who is not involved with your theory.

The control in experimentation

In most biological work, especially with people, we are never certain just how many variables we are dealing with, so we set up a control. For example, if we wished to see whether a drug helps cure a disease, it is not enough just to give the drug to people suffering from the disease and see how many recover, because we cannot be sure that the recoveries would not have occurred anyway. In cases like this two groups are selected, as nearly identical as possible. One group is given the drug, the other group a neutral material, 'the placebo', indistinguishable from the real tablet to the people being tested. The cure rates are then compared in the two groups.

To make the results more reliable the double blind technique should also be applied simultaneously. The dispenser of the drug should not know if what he dispenses is the drug or the placebo, nor should the patients know. If anything is said to the patients it should be repeated exactly for each person. The diagnosis of cure or not cure should ideally be done by another experimenter, and the results finally collated by a third.

How significant are your results?

No amount of mathematical processing can turn bad results into good, or put significance where there is none, but it is useful to put a measure to the reliability of your results as numerical data. Simple techniques for small samples are given in Part III.

References
1. Stubbs, J., *New Scientist*, 13 Jan. 1977, p. 59.
2. An interesting insight into the strengths and weaknesses of observational methods without statistical analysis can be obtained from Lorenz's general papers. *See* Nisbett, A., *Konrad Lorenz* (London: Dent, 1976).

Part I

The Growth, Size and Shape of the Body

1 Stature and weight

Introduction to Methods 1–8

The measurement of human height and weight are simple enough providing certain precautions are taken, but this simplicity belies the importance of these two measurements. By following height and weight changes normal healthy growth can be distinguished from abnormal growth, which may result from illness, both physical or mental, or may reflect nutritional restrictions. We now know without a doubt that height and weight has been steadily increasing in the Western world, and this raises the question, is there a limit to this process? We know too that height and weight criteria are used in many walks of life. Few professional footballers are over 6 ft 2 in. or less than 5 ft 6 in., indeed the average appears to be between 5 ft 8 in. and 5 ft 11 in., with weights of 11 to 12 stones respectively. London police have to be at least 5 ft 8 in. Some American police forces have a minimum height of 5 ft 10 in. Airline hostesses must not be over 5 ft 9 in. in most companies, while a ballerina of 5 ft 9 in. is not known. Some people have noted that American presidents, or certainly the last ten except Trueman and Carter, have been of greater than average height, but since UK politicians appear to be less than average height, future work on height and political leadership would have to be done. Not only are there obvious instances of the impact of height and weight on everyday life but these variables have a pattern throughout our lives, and some of these patterns are described below.

Some strange patterns have emerged lately. Swedish soldiers appear to be getting longer in the leg but their back length has not varied much during the last 100 years. Present day Japanese children outstrip their parents, probably because of improved nutrition, particularly in dairy produce, but they do not yet reach average European heights. The average heights of American college men and women have been increasing steadily at about 0.1 in. per year. Similar effects are seen in the UK. Obviously this trend must stop, but when, and at what height?

By applying the methods described in this section, you may be able to determine new patterns for yourself. The curious point is that very little is known about human dimensions because patterns have often been missed, and in many cases there is still a scarcity of information. The only way this paucity can be charted is by properly constructed experiments to discover trends and changes, as well as to give data for hypotheses.

Method 1. Measurement of height

(a) Standing

The precautions you have to take are quite extensive. Firstly, the floor must be level, and not give way under weight. Secondly, the measuring length must be at right angles to the floor. Thirdly, there must be some means of making sure you use a standard method. Some people slouch, some are erect, you *must* make sure you use a fixed method. You can get people to stand to attention! Or you can ask them to inhale fully. Try out different methods and you will see how heights vary considerably. As a guide, here is a standardized procedure. The person to be measured stands with feet (no shoes) about one inch apart. He or she draws themselves to full height without raising the shoulders, with hands and arms hanging relaxed. The measurer makes sure the rule is directly at the top of the skull and tries to make sure hair is not getting in the way. The breath is held at half-full inhalation (see Figure 1.1). Enter your result to the nearest millimetre, e.g. 165.3 cm rather than 165 cm.

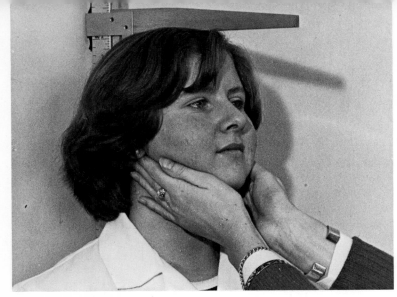

Figure 1.1 Measuring height.

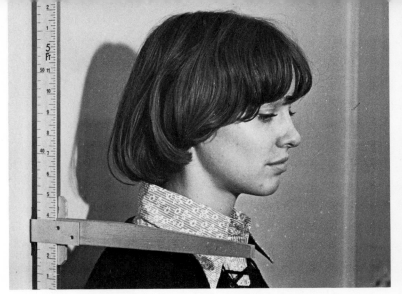

Figure 1.2 A method of measuring the 'height' of the human body.

(b) Supine

The same care is needed here, but you will have to make sure the person you are measuring is lying out straight, not curved. A startling fact for most people is to find that their standing height is less than their supine height.

Can you explain this? Do larger people shrink more or less than thin people?

Can you suggest any other 'heights' of the human body? (See Figure 1.2.) Can you work out how to measure them just using simple apparatus? Here are the results of a simple class experiment.

The subject lies on the floor, feet flat against the wall between two parallel 2-metre rules. A third rule is placed across the two rules level with the top of the subject's head and at right angles to the other rules. The measurement of the length is noted to the nearest centimetre.

Because the measurements in Table 1.1 were taken only to the nearest 1 cm, differences are probably exaggerated, but the tentative conclusion would be that there is a difference in the heights.

Table 1.1 Results of Method 1. Measurements are given in centimetres.

Subject	Height standing	Length supine	Difference
1	168	170	2
2	172	173	1
3	170	172	2
4	164	165	1
5	169	171	2
6	170	171	1

If the shrinkage when standing is real, then the obvious explanation would be compression of joints through gravity acting on body mass. You would need to convince yourself however, whether or not slouching or bad posture were not a cause in your experiments. If compression does occur you might expect people with heavy build (endo-types) to shrink more.

Method 2. Is height constant or does it vary during the day?

Several large scale studies have recently shown that height falls during the day. In a small scale experiment with women, all of age 20 years and on their feet most of the day working in laboratories and kitchens, the results given in Table 1.2 were obtained.

Observe that all six subjects show a fall of height during 7 hours. This information can be shown on a graph (see Figure 1.3).

Perform a similar test with a group of people. Can you give an explanation of your results? Is the average in height depression in your experiment bigger or smaller than the results shown in Figure 1.3?

Table 1.2 Results of Method 2. Measurements are given in centimetres.

Height measurement in morning (10 a.m.)	Height measurement in afternoon (5 p.m.)
170.0	169.0
160.0	159.5
165.0	164.0
171.5	170.5
162.5	162.0
169.0	168.5

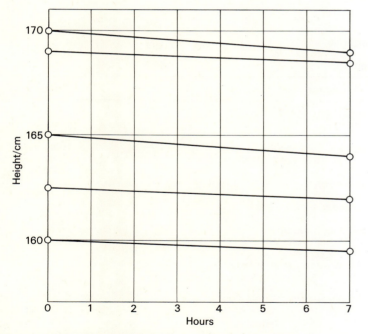

Figure 1.3 Variation of standing height (cm) with time (7 hours) of women during a working day. Five subjects were measured.

Method 3. Does weight vary during the day?

People vary greatly in weight from person to person, from the 100 kg of a heavy-weight boxer to the 50 kg of a jockey. The measurement of weight is important in many jobs, e.g. boxing and jockeys, and in grades in athletics, and in assessing obesity, and good nutrition. In precise studies weight is measured to the nearest ounce or about 25 gm, and when accuracy like this is necessary clothing must be taken into account (see Figure 1.4).

The weights of common articles of clothing are shown in Table 1.3 on the next page.

Weighing is usually done nude, or with a light under-garment, but other factors are important too. An adult person's bladder can hold 0.5 kg of urine, and so measurements are usually taken with an empty bladder. But are there other factors? Weigh yourself, and other people several times during one day. Can you detect differences? Can you explain them?

Human Measurement

Figure 1.4 Measuring weight.

Table 1.3 Weights of male and female clothing.

Female clothes	Weight (oz)
Bra, unpadded	1
padded	2
Pants	1
Pantigirdle, lightweight	3
medium/heavy	4–6
Petticoat, waistslip	2
full-length	3
Tights/socks	1
Blouse	4
Jumper, short sleeved	4
thin	4–6
thick	10–13
Skirt	8
Skirt if lined	10
Trousers, thin	13
thick	16
Jacket, lightweight	17
Cardigan, thin	8
thick	20–23
Belt	2–5

Male clothes	Weight
Pants	3 oz
Vest	4–5 oz
Socks	1 oz
Trousers, lightweight	16 oz
heavy, e.g. cord	20 oz
Shirt	6 oz
Jumper, thin	10 oz
thick	14–18 oz
Waistcoat, lightweight	6 oz
heavy	10 oz
Jacket, lightweight	24 oz
heavyweight	2 lb
Suits—2 piece	
Small–medium, lightweight	2 lb
medium	2 lb 8 oz
heavy	2 lb 12 oz
Medium–large, lightweight	3 lb 4 oz
medium	3 lb 8 oz
heavy	3 lb 12 oz
Tie	2 oz
Belt	5–10 oz

Method 4. Use of histograms to show distribution of heights

When data is collected in sufficient detail it is often useful to group the figures. Consider the data below. Heights of fifty males between 18 and 25 years were taken to the nearest centimetre. The investigators have grouped these results in cells of 3 cm. Note that some 'cells' have no values in them. If indeed the survey were taken further, all the cells would be filled, and it might be decided that it would be better to use a 1 cm cell. Once you decide on your cell, you pick out your values and count them up, obtaining a *frequency*. For example, in Table 1.4, the frequency of men found with heights 168–170.9 cm was 3.

Data can also be displayed in a histogram, as shown in Figure 1.5. Although the shape is irregular, it does appear that the distribution is normal, i.e., approximately equal numbers of 'cells' on either side of a

Table 1.4 Heights and frequencies for young men aged between 18 and 25 years, at the North London Polytechnic in 1972.

Height in cm	Frequency
159–161.9	1
162–164.9	1
165–167.9	0
168–170.9	3
171–173.9	8
174–176.9	9
177–179.9	11
180–182.9	9
183–185.9	2
186–188.9	6
189–191.9	0
192–194.9	1

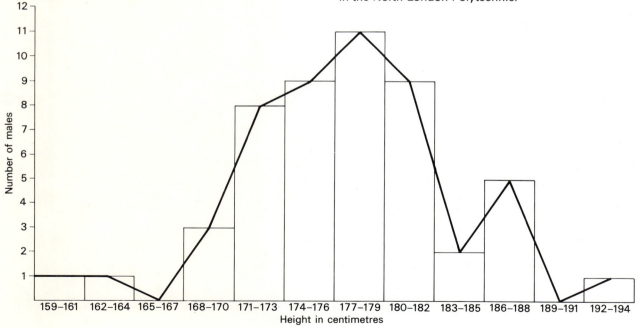

Figure 1.5 Histogram to show the heights of males between 18 and 25 in the North London Polytechnic.

Human Measurement

peak. It may be worthwhile to use a bigger cell, say twice the size shown (3 cm), and discover what histogram results. When large numbers of values are measured, frequencies show us the most common values.

You can now use this method for heights, weights (see Table 1.5 and Figure 1.6), or any measurements of a group of people. Find the frequencies for weights and heights in your class. (To produce results which could be permanent, that is of publishable value, measuring must be accurate.)

Table 1.5 Weights and frequencies for young men aged between 18 and 25 years, at the North London Polytechnic in 1972.

Weights in lb	Frequency
125–129	1
130–134	2
135–139	5
140–144	5
145–149	10
150–154	3
155–159	6
160–164	4
165–169	7
170–174	3
175–179	1
180–184	1
185–189	0
190–194	2

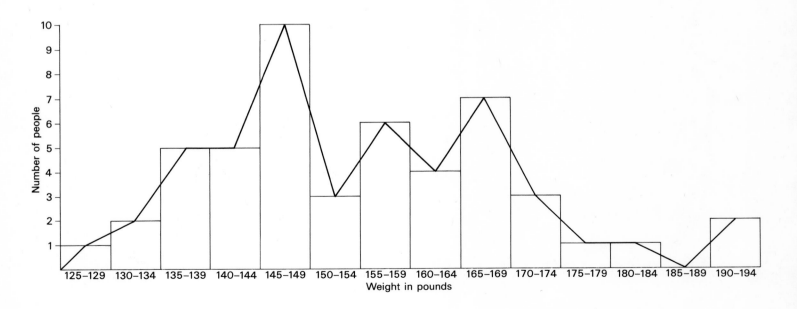

Figure 1.6 Histogram to show the weights of 50 males aged between 18 and 25 in the Polytechnic of North London.

Demonstration of Growth by Plotting Height and Weight Against Time

Method 5. The individual

In the graphs (Figures 1.7 and 1.8) the growth of a boy from his 317th week to the 535th week is shown. Observe how height increase in this period is nearly linear, whilst weight increase appears to accelerate. Both these features are shown by most children in this age group. Observe too how the curves have 'kinks' in them. This occurs because growth is seldom completely smooth, even minor illnesses can produce a drop in the rate, whilst weight varies during the day.

Figure 1.7 Graph of weight (kg) versus age (weeks) of a boy born on 11 February 1966 starting at the 317th week.

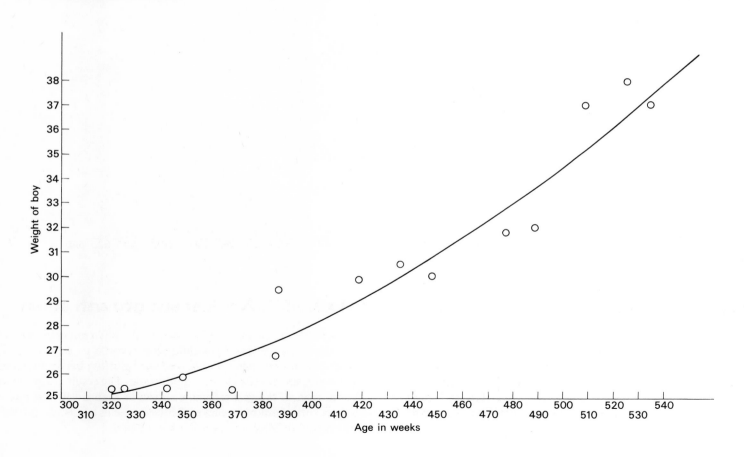

Human Measurement

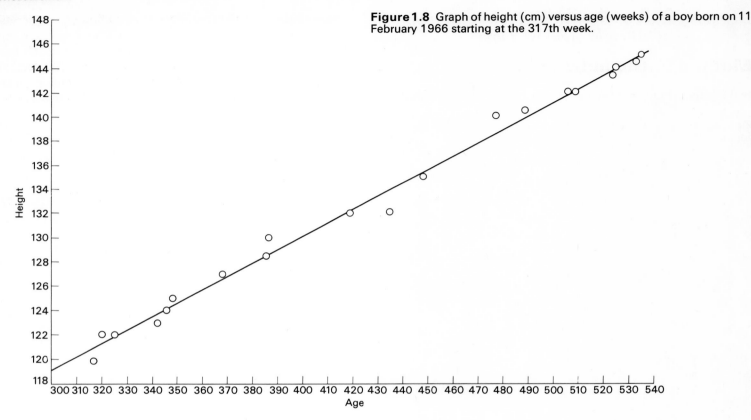

Figure 1.8 Graph of height (cm) versus age (weeks) of a boy born on 11 February 1966 starting at the 317th week.

Method 6. Groups

Individual growth patterns are still not so well studied as results obtained by taking heights and weights of different children at different ages. Studies following the same group of individuals provide a truer picture of growth. Try a longitudinal study in your school, and compare the results with those of the standard studies.[1] Also, try to follow growth by taking average heights and weights of different age groups of children.

Method 7. Adolescent growth spurt

Both height and weight begin to accelerate when puberty sets in. You can demonstrate this by maintaining records in any of the areas described in Methods 5 and 6, and then plotting height increases in, say, every six months or year against time. The result is a curve with a bump in it, of the general shape shown in Figure 1.9. The bump for girls occurs about two years earlier (see articles by Gray, and Tanner, mentioned in Method 6, for further details).

Figure 1.9 Typical height–velocity curve for boys.

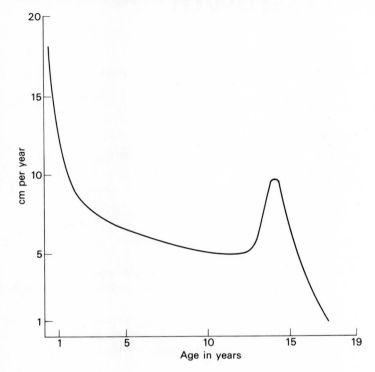

Stature and Weight

Method 8. Growth differences and sex

Plots of height and weight against age reveal that girls grow faster than boys between the ages of 8 and 14. Girls reach maximum height at about 16, while boys reach their maximum at 18.

To test whether someone is growing properly, plots of height and weight are made on standard charts which show not only the mean (50th percentile) growth curves, but also curves in what are called percentiles. A 10% percentile curve means that 90% of all children have larger values, and 10% have less. If an individual's growth is plotted on these standards then, providing he is in good health, his curve will tend to follow one of the percentile curves very closely. Sudden departures up or down from the growth curve will usually have a simple explanation, such as illness or changes in nutrition. These curves (compiled by Tanner *et al.*, loc. cit.), can be obtained from Creaseys Limited (Castlemead, Gascoyne Way, Hertford, Herts).

References
1. Tanner, J. M. and Whitehouse, R. H., *Archs Dis. Childhood*, 1976, **51** 170–79; see also an earlier study by Gray, G., *Scientific American*, 1953, **189** (4), 65–76.

2 Some Aspects of Human Growth

We still know very little about our growth; in particular we can only make a reasonably accurate estimate of a child's eventual height and weight, and the differences between children raised in different circumstances have not been examined in detail. We know that sound nutrition and parental love helps growth, but the effects of poor food and emotional neglect have only been touched on. We still know too little about the relative effects of these factors on different parts of the body's development, although we are certain that both height and weight are defective when emotional and nutritional needs, as well as exercise, are neglected.

Methods 9–12 examine some of these problems, and much useful data can be generated by careful study of these neglected areas. Sexual development is closely linked to a child's background, but again we have little data, while studies of the different ways in which various parts of the body grow as puberty is reached, and passed through, needs much more detailed study. Here you will find just how revealing simple, well planned measurements can be (see Method 12).

Method 9. Prediction of adult height

When the height of children is measured over the growth period, we can say with a fair degree of accuracy that at 8 years old, boys have reached 72%, and girls 77.5%, of their eventual height.[1] However this supposes that the child will receive adequate love, health care, affection, exercise and food, since otherwise growth will not meet expectations. Also a child entering adolescence a year earlier than average will not reach the predicted height for his chronological age, but for his biological age; in this case:

biological age = chronological age
  + earlier time of onset of puberty

Table 2.1 shows the accuracy of prediction for a boy, measured by the

Table 2.1 Expected adult height of a boy measured for three consecutive years.

Age (years)	13	14	15
Height (cm)	153.0	164.5	170.0
Percentage of adult height*	87.3	91.5	96.1
Expected adult height (cm)	175.3	179.8	176.9

*As predicted by Bayley, loc. cit.

author, who reached 177 cm at 18. At 13 years we could safely predict, in this case, that he would be tall enough to be a policeman, and too tall to be a jockey.

By recording height and age for boys and girls and following them to adult height, you can construct your own tables. Put the average height reached by your subjects at adult height as 100%, and calculate the percentage of this height for the average heights at three years, four years, and so on. This is a long term study but well worth it. You can also find the heights at different ages in a school for comparison.

Method 10. Growth and socio-economic class

Although there is a myth of the big 'working class' man, or the weedy 'professional man', human biology *does* differ. Children from higher socio-economic classes are taller and healthier than children from less privileged homes. You can check this yourself by finding mean heights and weights for children in different educational institutions. The cause of this is not genetic, since once diet and health are improved, as in Japan for instance, height increases.

Method 11. Onset of puberty and socio-economic class

In boys puberty can be identified when pubic hair appears. Other signs are increase of genital size and hair on the chin, with voice change. In girls, puberty is shown by an obvious change in shape (increase of hip relative to waist), pubic hair and breast growth, but the first menstruation or menarch is the ultimate sign that biological status has changed.

There are signs that both boys and girls are maturing earlier now than, for example, in the Victorian age. Can you make assessments, giving your criteria, of the onset of puberty in your school, or in a group of children? You should note that if you use hair growth criteria you will not necessarily get the same age as you would for other body changes, so you must always be clear as to your standards. Does, for instance, a girl's shape change before the appearance of pubic hair, or after? Are there any clear behavioural changes?

Method 12. Differential growth

Very approximately, the skeleton increases in weight from birth to maturity 20 times, the pituitary gland 5–10 times, while the weight of the nervous system increases by less than 5 times. This differential growth is also reflected in shape changes. Many questions on differential growth remain open because we still have insufficient data to answer them; for example: does the increase of the circumference of the head proceed at a faster or a lower rate in boys compared to girls?

Can you think of other forms of growth which could be investigated simply? Table 2.2 shows relative changes in growth of an adolescent girl, measured by the author, by monitoring circumferences. Note how the weight change has outstripped height change by a factor of 4.72; and how the hips and bust also have grown faster than height, while the waist has increased more slowly, thereby resulting in the typical female form.[2]

Table 2.2 Relative changes in growth of an adolescent girl; date of birth 11 July 1963, menarch (first period) April 1975. Unless stated otherwise, measurements are in centimetres.

	18 months before menarch	18 months after menarch	percentage change	percentage change / percentage change in height
Height	142.5	159.1	11.6	1.00
Weight (kg)	34.01	52.7	54.7	4.72
Neck	26.5	28.4	7.2	0.62
Head	52.5	53.6	2.1	0.18
Shoulders (circum.)	82.5	93	12.7	1.10
Chest (over nipples)	68.5	82.6	20.6	1.78
Chest (below bust)	63.5	71	11.8	1.02
Biceps	21.5	23.8	10.7	0.92
Forearm	19.5	21.8	11.8	1.02
Wrist	14.0	14.4	2.9	0.25
Waist	62	66.8	7.7	0.66
Hips	76.3	92.2	20.8	1.79
Pelvic width	18.3	20.5	12.0	1.03
Thigh	46.5	53.8	15.7	1.35
Knee	29.5	35.0	18.6	1.60
Calf	28	31.5	12.5	1.07
Ankle	19.2	21.0	9.4	0.81
Leg length (inside)	65	78	20.0	1.72

Percentage change = $\dfrac{\text{measurement taken after menarch} - \text{measurement taken before menarch}}{\text{measurement taken before menarch}} \times 100$. For details on methods of measuring, see Chapters 1 and 3.

Human Measurement

Introduction to Methods 13–15

The Renaissance artists were convinced that mathematics and human anatomy were inseparable. Leonardo da Vinci inscribed the human figure within a circle and simultaneously within a square. Dürer tried to construct his bodies on precise mathematical formulae (see *Dresden Notebook*). Today we would not say a person whose height was $6\frac{1}{2}$ times their head length was a 'peasant-type', but in advertising drawings legs are lengthened out of real proportions to give the intended lean, fashionable, look.

Our bodies have proportions of course, but they are also asymmetric. Arm span and height are very close, as da Vinci's circle and square man suggests, but the left and right sides of our bodies are not identical, the heart for example, is to the left, while, functionally, left and right sides differ in handedness and development. The symmetry is nevertheless a strong feature, and suggests an ordered growth and development from the fertilized egg we all come from. We still do not know more than a fraction of the quantitative relationships between various parts of our bodies. These must signify not only our genetic inheritance, but the effect of nutrition and health upon our development: in a word, our growth.

Method 13. Correlation of arm span and height

Measure height as described in Chapter 1, and arm span as shown in the photographs (Figures 2.1 a–c). In Figure 2.2 the arm span data were taken on full exhalation. Observe how there is a distinct trend for larger heights to follow longer arm-spans. Is the trend shown as clearly with children under 16 years?

A statistical treatment of correlation is given in the appendix. Apply the methods described to the data in Figure 2.1.

Figure 2.1 (opposite) Measuring arm span. Note how an accurate measure is obtained using a piece of chalk sharpened to a wedge, giving a clear straight line for the tape to be put against.

(a)

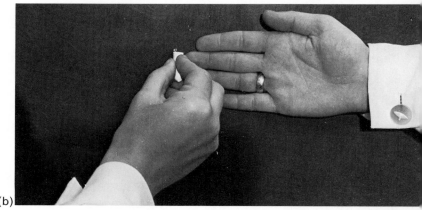
(b)

(c)

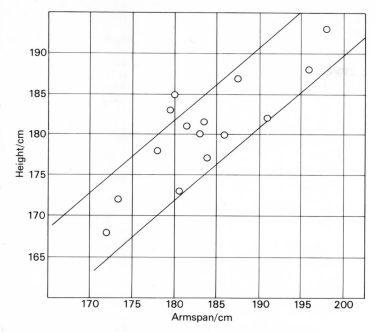

Figure 2.2 Height as a function of armspan for young men.

Some Aspects of Human Growth

Table 2.3 Average asymmetry ratios of the circumference of right limbs to left limbs of young women at the Polytechnic of North London.

	Survey 1	Survey 2
Ankle	1.02	1.006
Calf	1.06	1.02
Thigh	1.05	—
Biceps	1.09 (flexed)	1.04 (relaxed)
Wrist	—	1.02
No. of subjects	15	8

Survey 1. Harris, A. B., *Nature*, 1976, **261**, 10.
Survey 2. Whalley, J., *et al.*, *Nature*, 1976, **262**, 253.
(See methods for circumference measurement in Chapter 3.)

By measuring weights of upper limbs and various muscles of corpses, Chibber and Singh[3] also found differences between left and right limbs, but there were individual differences in the various muscles. They remarked 'we made the rather surprising discovery that virtually nothing was known about the anatomical aspects of handedness'. This remains as true now as it was in 1972.

Investigate asymmetry by measuring circumferences.

Asymmetry

Method 14. Demonstration of asymmetry of limbs by circumference measurements

By measuring the circumferences of the limbs on the right hand side of the body and the corresponding values on the left-hand side, and finding the ratios right : left of the circumferences, an asymmetry is revealed if the ratio differs from unity.

For young women at the Polytechnic of North London the results given in Table 2.3 were obtained.

Method 15. Handedness

As a first hypothesis we can assume that a right-handed person will have larger muscles (therefore circumference) on the right arm, since we know that muscular use increases size of muscles and disuse leads to atrophy. But what of the leg? Are we 'legged' as well as handed? The results in Method 14 suggest we are. However, there is evidence that the human foetus shows a preferential growth in the right-handed limbs, indicating asymmetry is programmed during foetal growth. The embryo does move in the womb, so this *may* be the result of preferential use.

Socially, children who show a preference to use the left hand, are still too often taught to change to the right in writing and use of knife and fork. If handedness is genetically determined, or results from preferential movement during foetal life, the practice would appear, at least, ill-advised. Ambidexterity is also observed. Leonardo da Vinci could write simultaneously with both hands, even producing a mirror image of his right script in his left script.

Determine what percentage of children in a class write with the right hand. What percentage have been taught to do this against their natural inclination? Using the grip meter, determine the percentage of strength in the left hand compared to the right.[4]

References
1. Bayley, N., *J. Pediat.*, 1956, **48**, 187–94.
2. An excellent text on growth is Sinclair, D., *Human Growth after Birth*, (London: Oxford University Press, 1969).
3. Chibber, S. R., and Singh, I., *Acta Anat.*, 1972, **81**, 462.
4. A recent discussion on human asymmetry appeared in *Nature*, see Harris, A. B., 1976, **261**, 10; and Mittwoch, U., 1976, **261**, 364.

3. Analysis and Classification of Physique by External form

Introduction to Methods 16–21

People have been interested in classifying the shape of the human body since recorded history, but it is a subject still in its infancy. We expect boxers to have tough muscular physiques, sprinters to have muscular legs, and we have ideas of what kind of person a physique might imply. Cassius was lean, and Caesar feared him; Pickwick is jolly and portly. We do not expect Tarzan to be fat or thin; while we do know that soldiers and policemen, athletes, nurses, all have to have durable bodies.

Look around your school or college. Are people all the same? No, they are all different. But how could we go about describing a physique? A start might be to say Tarzan type, beauty queen type, female gymnast type and so on, and this could be a good beginning because you would be using the idea of form being some how related to function, however roughly.

Hercules is a *mesomorph*, Billy Bunter is an *endomorph*, the 'human skeleton' of the circus is an *ectomorph*. We owe these terms to Dr William Sheldon, an American scientist, who developed the first really useful method of classifying physiques. His methods are constantly being modified and improved by other workers, but his original terms have passed into our language.

Dr Sheldon began very scientifically. He examined thousands of physiques and photographs of physiques and found three extreme, rare, physiques. The very thin person, with hardly any depth or width, very thin angular bones, with muscle and almost no fat. This physique he called the ectomorph. The heavy boned, lean, but muscularly massive physique he called the mesomorph. The heavy, barrel-shaped, fat person he called the endomorph. His next step was very perceptive, because he suggested each of us is made up of differing mixtures of these three physiques.

Unfortunately Dr Sheldon's method is not without its critics in detail, but his general ideas are accepted as sound. In the following sections there is a much simplified and modified method of assessing these components; also the method given here can be used for either sex, and should be referred to as 'body-typing'. You can only do this by trial and error, rather than just reading about it. Study the photograph and diagrams (Figures 3.1–3.10) carefully, and compare them with people you know.

A scoring system for body-typing

The method described below permits a very rapid assessment of body-type in three physique components. The following lists for endomorphy, mesomorphy, and ectomorphy are so designed to make your attention move up and down the body during the assessment, thereby giving a unified result. A standard procedure is to view the physique frontally, laterally, and posteriorly. A perfectly expressed example of endomorphy at level 3 would score 3 for each attribute; in practice this is never found for different parts of the anatomy will score differently. People whose scoring lies within a narrow range are physically *homogenous*, and this is quite rare.

Inspectional Standard for Body-typing
Method 16. The first component, endomorphy

There follows here a complete check list of endomorphic characteristics. People with all these characteristics at a high level (7, 6, 5) are very rare indeed. It is *not* unusual to score low in some, high in others,

Human Measurement

in the list. Study the photographs and illustrations before proceeding. The score in *endomorphy* is arbitrarily fixed at a *maximum* of 7 and a *minimum* of 1.

On assessing endomorphy, you award 7 points when you think the characteristic is shown in an *extreme* form. If you think it is entirely absent award 1. Score as follows:

7—extreme; 6—strongly shown; 5—moderately strongly shown; 4—half present, half not present (you will not be seriously in error if you use your notion of average as score 4); 3—weakly to moderately shown; 2—very weakly shown; 1—absent.

This approach to scoring is also used for the other two components, mesomorphy and ectomorphy.

How to use the check list

Fix in your mind the attribute described, and for that attribute score the assessed person on the scale 1–7. It is useful to ask yourself, 'is this average?' If the answer is '*yes*', score 4; if '*no*', then proceed up and down the scale as the case may be. When you have dealt with that attribute, pass on to the next. When you have done them all, *add the scores and divide by the number of attributes you have checked*. This gives the *endomorphic rating only*.

You will usually get a fractional result, which should be written as the nearest whole number, e.g. 3.8 should be written as 4.

Figure 3.1 Endomorphy rating of 5.

Figure 3.2 Endomorphy rating of 6.

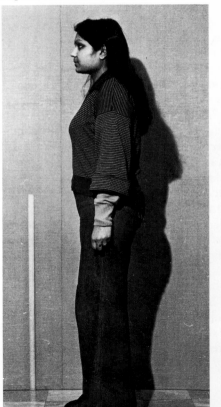

Figure 3.3 Endomorphy rating o

Scoring list

1. Roundness and softness of body.
   Width and depth tend to be the same:
2. In trunk (as deep as it is broad).
3. In head (as long as it is wide).
4. In limbs (thighs round, ankle and wrist round, calf round).
5. In neck (round).
6. The trunk appears to be the main part of the body, in volume, the limbs very secondary.
7. The abdomen is larger than the chest.
8. The upper parts of the limbs very much bigger than the extremities.
9. Pronounced ham-shape of upper arms and upper legs.
10. Shoulders high and square, but soft.
11. Short neck.
12. Head round.
13. Face round.
14. Head is large.
    Muscles do not show clearly, contours smooth:
15. No clear calf muscle.
16. No clear shoulder muscle (deltoid).
17. No clear neck to shoulder muscle (trapezius).
18. Abdominals not observable.
19. Limbs short and tapering.
20. No clear jaw line.
21. The bones of wrist and ankle, knee and elbow, small and round, they do not project.
22. On feeling, bones are smooth and round, especially of joints, of pelvis, and shoulders. Collar bone not clearly seen.
23. The back is straight from side view, the spinal column without *S*-shape (high endomorphs have straight backs).
24. The abdomen is well packed, tightly packed, with organs.
25. The waistline is high and not clearly shaped.
26. The chest is as wide at its base as across the upper part.
27. The greatest width below the waist is high, rather than across at the crotch.
28. The rib cage is smooth-shaped, like a flat bow, at the sternum.
29. The lower ribs lie horizontally in side view.
30. The male has pseudo breasts (male endomorphs have a distinctly female shape).
31. Skin is soft, smooth and velvety.
32. Ears lie flat.
33. Nose soft and not projecting.

In general, you may assess each characteristic when the subject is nude, but quicker assessments can be made without full nudity which do not differ substantially from the more prolonged assessment. However, when not completing the total list, you must make sure you cover the major parts of the body, enumerate your total, and divide by the number of scorings you have made.

Method 17. The second component: mesomorphy

Mesomorphy means physical ruggedness and strength. In men it means big bones, big joints, strong muscles; in women the muscularity is obviously not so pronounced but is still there. Remember when scoring that 4 for a man in mesomorphy is not the same as 4 for a woman. Four is the mid-way point for men and women, but obviously the scales are different. A score of 7 in men and women is very rare.

Scoring list

1. Square strong body.
2. Muscles clearly defined.
3. Trunk broad and deep, but not rounded.
4. Forearms approaching upper arms in thickness.
5. Legs with massive well defined calves.
6. Wrists very strongly made.
7. Ankles very strongly made.
8. Strongly made and strong-jointed feet and fingers.

Human Measurement

Figure 3.4 Mesomorphy rating of 5½.

Figure 3.5 Mesomorphy rating of 5.

9. Joints strong and large.
10. Chest is deep and broad, larger than abdomen.
11. Shoulder muscles are large and prominent.
12. Neck muscles are large and sloping.
13. Wide and deep pelvis.
14. Bones massive.
15. Strongly made head.
16. Square jaw.
17. Thick muscular neck.
18. Large, strongly muscled shoulder blades.
19. Straight upper back with prominent curve in at the middle of back.
20. Prominent muscular buttocks.
21. Tummy muscles clear and rippling.
22. Ribs are thick and strongly made.
23. Low waist, wider than deep.
24. Muscles of calves, thighs, arms, and chest very clearly marked and obvious.
25. Coarse skin, tanned easily, elastic.
26. Conspicuous pores.
27. Ligaments and tendons are thick and strong.
28. Joints are tightly sewn together.
29. Facial lines deep not finely etched.
30. Strong bony nose.
31. Thick firm lips.
32. The face is strong and rugged, not bland as in endomorphy or sensitive and fragile in ectomorphy.
33. Strongly boned skull, square.

Method 18. The third component: ectomorphy

Ectomorphy means fragility of bone structure, thinness of muscles, little depth in body, little width, little roundness of body shape, thinness and shallowness.

Analysis and Classification of Physique by External form

Figure 3.6 Ectomorphy rating of 5½. Observe roundedness of form giving an endomorphy rating of 4, despite lack of bulk.

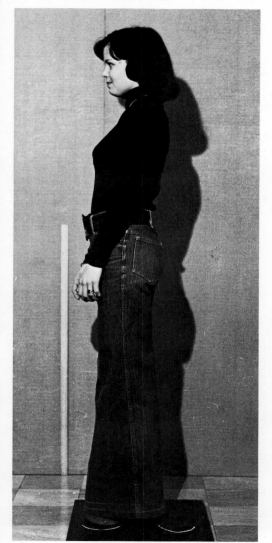

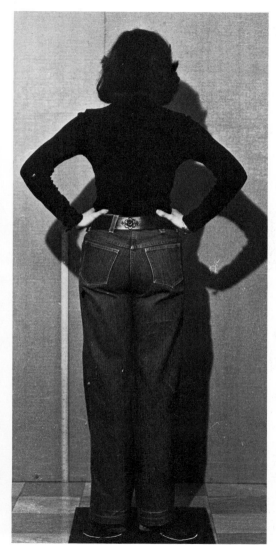

Human Measurement

Figure 3.7 Ecto-meso-type.

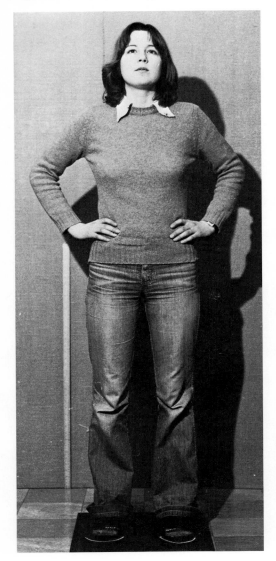
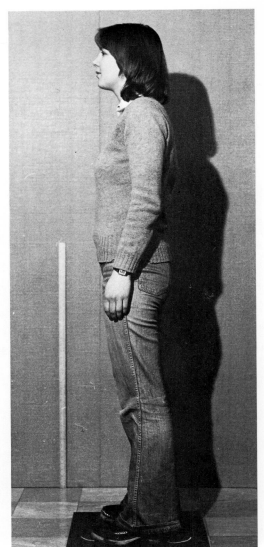
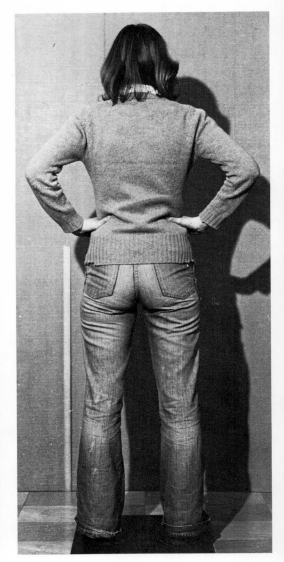

Scoring list

1. Long thin body.
2. Thin bones.
3. Stringy muscles.
4. Very thin on side view.
5. Drooping shoulders (slouched).
6. Long limbs in comparison to length of trunk.
7. Flat, shallow abdomen (but see 15 in cases of overweight).
8. S-shaped spine but deeper bend than S-curve of mesomorphy which is more rounded and powerful looking.
9. Chest narrows in depth very much from the nipples upwards.
10. Prominent delineated ribs.
11. Narrow shoulders.
12. Little muscle on body.
13. Marked hollow behind collar bone.
14. Shoulder blades stick out.
15. Weak muscles of abdomen results in distended tummy *below* navel.
16. Extremely weak looking upper arms and thighs.
17. Long forearms and shins in comparison to upper part of limbs.
18. Fingers and toes fragile and thin, joints are not prominent but little and pointed (prominent joints indicate mesomorphy).
19. Long thin neck.
20. Neck bends forwards.
21. Very weak muscles joining neck to shoulder.
22. Larger forehead than face.
23. Facial features small, pointed.
24. Pointed nose, sharp thin nose bridge.
25. Weak bones of jaw, often receding.
26. Thin delineated lips.
27. Fragile looking skull, not rounded.
28. Skin thin and often scaly.
29. Skin wrinkles easily.
30. Skin does not tan easily.
31. Little elasticity in skin (a pinch takes time to return to original position).
32. Abdomen feels slack.
33. Joints loose, tendons thin.
34. Joints are pointy although not prominent.
35. Hair is fine.

Method 19. Body-typing

Now that three components of physique can be measured, we see that any particular body can be given a number in this order: endomorphy rating, mesomorphy rating, ectomorphy rating. A little thought reveals that there would be $7 \times 7 \times 7$ theoretical possibilities, except that it is difficult to see how anyone could score 7-7-7 or 1-1-1. Nevertheless a large number of possibilities exist. Sheldon found nearly a hundred different physiques, but there are difficulties in dealing with so delicate a division, and so seven different body-types which cover any population are described in Table 3.1.

Table 3.1 Conversion table of endomorphy, mesomorphy, and ectomorphy ratings into the seven main body-types.

Type	Component scores between		
	Endomorphy	Mesomorphy	Ectomorphy
Balanced (Figure 3.8 (a), (b))	3 and 5	3 and 5	3 and 5
Ecto-meso	1 and 2	4 and 5	3 and 5
Endo-meso	4 and 5	4 and 6	1 and 2
Endo-ecto	3 and 5	1 and 2	4 and 5
Meso (Figure 3.8 (c), (d))	1 and 3	5 and 7	1 and 3
Ecto (Figure 3.8 (e), (f))	1 and 3	1 and 3	5 and 7
Endo (Figure 3.8 (g), (h))	5 and 7	1 and 3	1 and 3

To classify any physique refer to the male or female photographs and to the check list to score the component ratings. Once you have the ratings refer to Table 3.1 to find the type of physique.

We have body-typed several hundred women at the Polytechnic of North London. Below is given an example of the common balanced physique.

3.3-4.4-3.9

Human Measurement

Figure 3.8 Standard body-types.

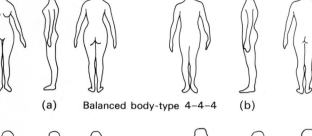

(a)   Balanced body-type 4–4–4   (b)

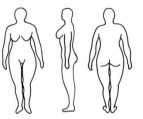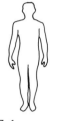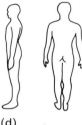

(c)   Meso-type 2–7–1   (d)

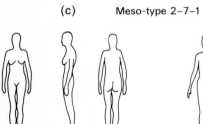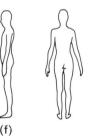

(e)   Ecto-type 1–2–7   (f)

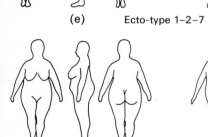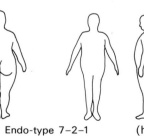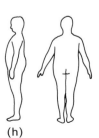

(g)   Endo-type 7–2–1   (h)

Figure 3.9 Endo-meso-type.

Analysis and Classification of Physique by External form

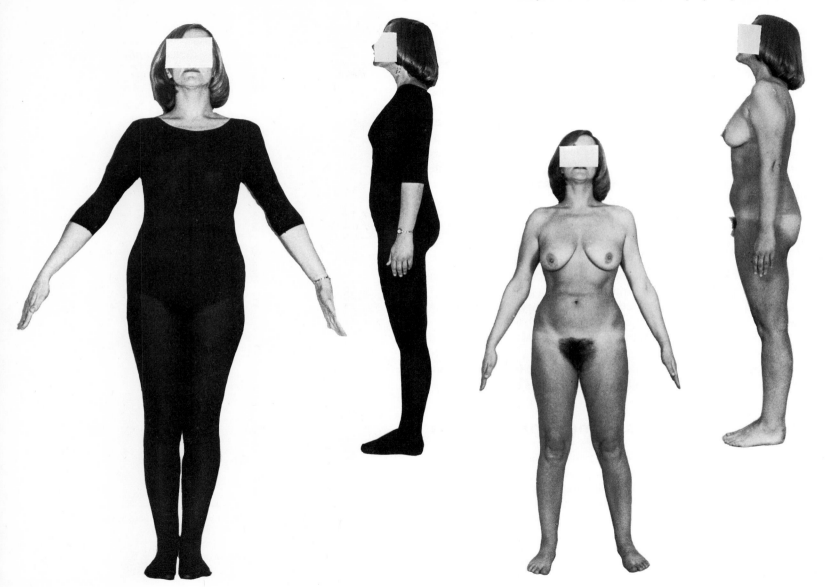

Human Measurement

These values are the average component ratings. When they are corrected to the nearest whole number, we obtain

3-4-4

Reference to the conversion table shows that 3-4-4 is classified as a balanced physique. A completely worked out example is given in Table 3.2 (see Figure 3.10). In the example detailed in Table 3.2, the average, corrected to the nearest whole number, is 2-4-3. Reference to the conversion table (Table 3.1) indicates this as an *ecto-meso-type*.

The body-typing procedure described can be used for a variety of investigations, and any you can think of will have a good chance of being original since so little work has been done, especially on female physiques. There are some tables given here of what results have been obtained, but try to apply the method to your own environment.

Suggestions for future study

1. Is it desirable to use measurements instead of check lists? Can you begin a set of standards for your age group?
2. Can you assess sexual attractiveness or beauty in body-typing terms?
3. What physique types appear most socially successful, or is there no such correlation?
4. Can you discuss any physique trends in the pupils of your school or college; are there any trends in the physiques of the staff?

Method 20. Self assessment of physique

Use Methods 16 to 19 to determine your physique type. Does it agree with someone else's assessment of you?

When examining yourself use a full length mirror, view yourself from behind and sideways using another mirror, and fill in each trait with its score.

Table 3.2 Score sheet for body-typing of female subject shown in Figure 3.10 (page 27).

Trait	Endomorphy	Mesomorphy	Ectomorphy
1	3	4	4
2	4	4	4
3	2	4	4
4	4	4	4
5	2	2	2
6	1	5	3
7	2	4	3
8	3	4	3
9	2	4	5
10	1	4	3
11	2	4	2
12	1	4	3
13	1	4	6
14	2	4	4
15	4	4	1
16	3	4	1
17	2	4	2
18	2	4	3
19	1	4	4
20	2	4	4
21	1	4	4
22	1	4	4
23	1	4	4
24	2	4	4
25	1	4	4
26	1	4	2
27	1	4	2
28	2	4	2
29	2	4	4
30	—	4	4
31	2	4	4
32	1	3	3
33	1	3	4
34	—	—	4
35	—	—	4
Mean body-type	2	4	3

Reference to Table 3.1 reveals the physique as an ecto-meso-type (see appendix for full statistical analysis of this physique).

Analysis and Classification of Physique by External form

Figure 3.10 An ecto-meso physique.

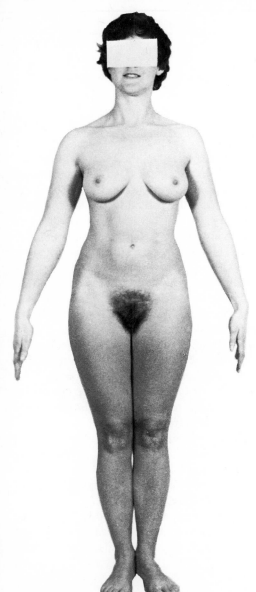

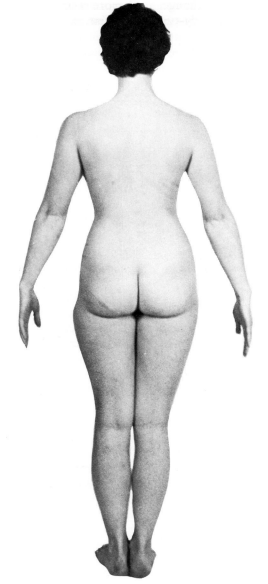

Although much more work needs to be done on the actual numbers of body-types that exist in any given population, we are reasonably certain that some are more common than others. Table 3.3 shows the approximate percentage of each body-type in an average population of men and women.

Table 3.3 Percentage representation of the main body-types in an average population.

Body-type	Percentage body-type	
	Male	Female
Balanced	41	27
Ecto-meso	10.5	25
Endo-meso	12	19
Endo-ecto	3	—
Meso	18	10
Ecto	10	11
Endo	5.5	—

The absent entries in the women's column are rare Body-types (see appendix).

We have studied the physiques of mongols (i.e. people born with an extra chromosome). They score high in endomorphy, low in mesomorphy and ectomorphy, and are classified as endotypes (6-2½-2). A rare condition (2 in 700 births) is the appearance of a further X chromosome in males, giving XXY instead of XY. One such individual, studied on film, revealed the body-type 5½-2-5½.

The Ballet Rambert, London, kindly gave us permission to classify some members of their company. The results were as shown in Table 3.4. The tallest girl was 5′ 6″, while the men were closer to average stature. Note how the female dancers are less typed than the men who are predominantly ecto-meso-types. Since in modern dance male dancers lift and manipulate their partners, a tendency to meso-typing would be expected, and this is confirmed by the analysis above. The female dancers are lower in endomorphy and higher in ectomorphy than college women (see appendix); this feature, with their small stature (average 5 feet), implies lightness, again confirming prediction, but observe subject 5, a strong ecto-meso-type.

Table 3.4 Classification of some members of the Ballet Rambert.

	Women			Men		
1.	4	2	5	1½	5	5
2.	1	2	5	2½	4	4
3.	3½	2	4	2½	5	3
4.	2½	2	5½	2	4½	4½
5.	2	5	5	1	5½	3
6.	1½	3	5	2	5	5
7.	2½	3	5½	3½	4	4
8.	2	5	5	2½	5	5

Although several assessors can agree on their results using the check-lists and the photographs as guides, some measure of objectivity is also required. It is commonly agreed that Sheldon's measurements of widths of points of the body do not give consistent results. However there are very good structural reasons for using circumferences as criteria in body-typing. Useful measurements are circumferences of chest, waist, hips, thigh, calf, bicep, neck, forearm, ankle, and wrist.

We should expect that meso-types will have bigger chests, thighs and calves than, for instance, ecto-types, because meso-types are defined as people with large bones and muscles. However we must always be careful to compare physiques at the same height.

By measuring a selection of circumferences of different body-types investigate the ranges of values obtained. You should note that measurements alone can never displace the photographs because, for example, both endo-types and meso-types will have large circumference measurements, but for different reasons. The endo-types have large deposits of fat, the meso-types large muscles.

In the studies at the Polytechnic we found that endo-ecto-types had almost exactly the same measurements as ecto-types, but were very different to look at (see drawings and photos).

See if you can distinguish body-types by measurement of circumferences using a non-stretch fibre glass tape.

A rewarding way to begin is to compare different body-types of the same height, for example ecto-types and endo-types. You will soon find that most bodies are mixtures of the components. A person may have a high score of ectomorphy in the legs, but have powerful shoulders and chest.

Method 21. Quantitative scales for body-types

Sheldon hoped he could eventually have a numerical classification system, and he tried comparing widths of the body, but unfortunately the results were very disappointing, even when height was allowed for, as shown in Table 3.5. Here you can see how the ratios overlap in different physiques, and so cannot be used to identify the endomorphy score because a single measurement can have several interpretations.

Table 3.5 Comparison of body-width ratios with endomorphy.

Ratio of back width over height	Endomorphy
16.8–21.6	1
16.8–22.0	2
18.4–22.4	3

(Adapted from Sheldon, *Varieties of Human Physique*.)

The reason for this is that a person's physique varies over their whole bodies, in other words they are heterogenous, and so any manageable set of measurements on a real body are going to fit several categories. However, mesomorphs will have bigger measurements than ectomorphs, and so measurements can help if used as a guide. This is perfectly acceptable because the aim of a classification system is to identify differences so that groups of people can be split into smaller groups. It is a question of how fine you want the subdivisions to be.

The biological finding in human physique is that it is continuously variable, and that the few parameters which have been measured on a sufficient scale reveal a normal distribution; but the impact of concepts of type on mere measurements reveals startling insights. If one measures tens of thousands of adults without reference to sex one merely gets a set of figures, but if one classifies the data at the *onset* according to a gross morphological typing, and the classification is based on possession or absence of genitalia which one does not have to measure to identify, then immediately important properties ensuing from this gross morphological typing appear; e.g., men are taller and heavier than women. But the heights and weights of men cannot be used to identify men as distinct from women. The strength of body typing lies in categorization by visual inspection. From this numerical data can follow.

By purposely picking out good physical examples of various body-types, representatives scales can be established. In Table 3.6 the *ponderal index* (PI)* clearly distinguishes endo- and ecto-types. An ecto-type with a PI of 38 cannot exist, nor a meso-type of PI 46. Finer differences exist, as you can see, and in general the PI is a good first criterion for body-typing, but since it does not distinguish between fat and muscle, or bone, in contributing to weight, these differences are not shown by PI alone. The ponderal indices of five very muscular physiques of professional body-builders was 38.4 ± 0.29,[1] while three meso-types without hypertrophy of the muscles had indices of 41.2, 43, and 43.6.[2] These findings confirm those of Table 3.6.

The ponderal index is a good check when dealing with very short

Table 3.6 Ponderal index for men and women (height in cm, weight in kg).

Body-type	Mean Ponderal Index	
	Men*	Women†
Balanced	43	42.7
Ecto-type	46	44.9
Endo-type	38	38.5
Meso-type	41	42.5
Endo-meso-type	39.5	41.5
Ecto-meso-type	43.5	43.0
Endo-ecto-type	45	45.0

*Selected from Sheldon's Atlas of men, and students of PNL.
†Female students PNL selected for homogenety.

*PI = height (cm)/cube root of weight (kg).

people, since one tends to score their bone sizes and general body bulk as smaller than they are. As it is the relative values to height that denote build, one can check one's impression of a physique by finding the index, and then from that value calculate the weight at average height for the group to which the individual belongs. A subsequentially lower than average and higher than average value can, therefore, act as a guide to one's general assessment. Similar remarks apply to very tall people, but in reverse, since even ectomorphic people at 6' 5" *appear* to have large joints and may be scored too high in mesomorphy. If the build is large, the tendency is to score too high in mesomorphy. The ponderal index can help guard against this type of mistake.

Circumference measurements

Providing the skinfold measurements are less than 20 mm, circumference measurements quickly reveal the type of physique, measuring as they do bone and muscle dimensions. The values in Table 3.7 were obtained at the Polytechnic of North London (PNL) by measuring circumferences on good, ideal, examples of the various body-types and idealizing the means. For position of measurement, see Figure 3.11. The mixed body types show such variation, as they do with widths, that a rule of thumb is to interpolate between the various types. Endo-meso-types and endo-types can show more massive measurements than these ratios reveal.

The circumference data per unit height, both measured in the same units, for three young male meso-types is shown in Table 3.8 when it will be seen the values agree with those in Table 3.7.

Quantitative scales for women

The constitutional literature is very poor in data relating to female physiques, and I hope this section serves as encouragement to undertake this work. One of the few studies giving standards was done just after World War II,[3] where height and weight were listed on 175 college women, but no gross morphological results were published. To redress this imbalance, I have presented here the mean

Table 3.7 Standard circumference ratios for men. Ratio of circumference to height (measured in cm).

Body-type	Balanced	Meso-type	Ecto-type	Endo-type
Chest	.55	.59	.49	.60
Waist	.42	.43	.40	.51
Hips	.53	.56	.50	.60
Upper thigh	.31	.33	.29	.36
Ankle	.13	.13	.11	.14
Wrist	.10	.11	.09	.11
Neck	.20	.22	.19	.23
Calf	.20	.21	.19	.23
Bicep (relaxed)	.18	.20	.13	.21

Figure 3.11 Circumference measurements. See Tables 3.11 and 3.12 for more details of measuring convention.

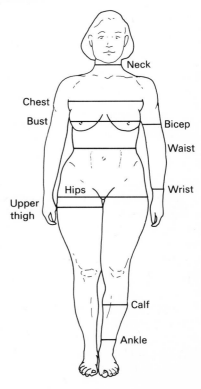

Table 3.8 Circumference ratios for 3 young mesotypes.

Subject	1	2	3
Somatotype	2-6-3	2-5-3	3-5½-2
Body-type	meso-type	meso-type	meso-type
Age (years)	20	21	18
Upper arm	.16	.15	.16
Thigh	.30	.29	.32
Calf	.21	.20	.22

Calculated from data of Tanner, J. M.[2]

values of a pilot survey on female body-types. We picked out homogenous physiques; people with obvious mixtures of components in different parts of their bodies were not included, nor were obese physiques.

There are significant differences between some of the body-types (e.g., ecto-types and endo-meso-types) even though we can hardly expect much difference when values are merely averages, but width and circumference criteria are useful when checking body-type ratings, and are best used as supportive material for the routine body-type physical assessment.

Ponderal index. The value of PI in assessing body-type for women is subject to the same qualification as for men, see Table 3.6.

Circumference standards for women (Table 3.9). Means were calculated from measurements of selected persons showing homogenous physiques of type shown. No attempt has been made to indicate ranges, but the tables can be used to support body-typing assessments using the trait lists and the diagrams, and for checking gross errors, for example confusing the massive endo, endo-meso, and meso physiques with the balanced, ecto, and ecto-meso physiques; the figures cannot be used to distinguish on their criteria alone endo-ecto and ecto-types, since these have very similar ratios, nor to distinguish endo-meso and meso-types. Since these figures refer to selected physiques they merely illustrate the type of physique; ecto-meso-types in particular show wide variations of measurements as a group since dysplasia (i.e., heterogenous development) is very common. Most real physiques are mixtures of types, and the possession of the above measurements does not imply that the scoring for the rest of the body will fit the same type. For this reason the check list of traits must remain the standard.

Circumference values must be obtained on the living physique, and whereas many can be measured with clothes, many cannot, as shown in Table 3.10.

Table 3.9 Mean body ratios of circumference to height for women of homogeneous physiques.

	Balanced	Ecto	Meso-ecto	Endo	Meso	Meso-endo	Ecto-endo
Bust	.54	.51	.52	.60	.54	.58	.51
Waist	.42	.40	.42	.49	.41	.47	.40
Hips	.58	.55	.56	.64	.58	.61	.55
Wrist	.09	.09	.09	.09	.10	.10	.09
Ankle	.13	.12	.13	.13	.13	.13	.13
Upper arm	.14	.13	.14	.18	.16	.16	.13
Calf	.20(5)	.19	.21	.23	.22	.22	.19
Upper thigh	.33	.31	.33	.36	.35	.35	.31

For precise position of measurement, see Figure 3.11.

Human Measurement

Table 3.10 Circumference measurements without and with clothes (in centimetres).

Measurement	Nude	Clothed	Difference
Ankle	22	22	0
Calf	36	36	0
Knee	40	40	0
Thigh	58	58	0
Hips	94	97	3.0
Waist	65	67	2.0
Chest			
Below nipple	73.5	74	0.5
On nipple	89	89	0
Above nipple	86.5	91	4.5
Neck	31.5	31.5	0
Head, round	57	57	0
Ear to chin	64	64	0

You can test the tables by assessing living physiques via the trait lists of Methods 17, 18, 19 and 20 and then taking measurements. They should agree. As a preliminary, assess the physiques shown in photographs where different degrees of nudity are purposely shown (see Figure 3.9). Do you get the same body-type from the clothed and the nude? Do you get the same body-type using the quantitative width scales?

We still do not know enough about the various proportions of body-types of women, and we know little about children. You could find out the proportions in your own school, or any selected group, and use the methods in Part III to make sure your results were of a publishable standard.

Angular description of physiques

Widths, circumferences, lengths of the body can all be used for classifying, on different systems, the human physique. An entirely new method can be constructed by taking photographs, in standardized pose, and measuring various angles. For example the more endomorphic figure on the right in Figure 3.12 has a shoulder angle of 140°, while the angle of the more ectomorphic figure is 120°. Other angles of interest would be the line of hips, and the line of the rib cage, both laterally and from the front and back. Take photographs of measured angles to see the variations that occur in different physiques.

Body-indexing in rapid survey work

As Professor Mayer of Harvard points out[4] different classification methods are used for different purposes, and we choose the most appropriate. Ingenious attempts have been made to make somato-typing quantitative, but by its very nature, this is impossible. One method involves very time consuming measurements and abstruse calculations.[5] Against such a background it is little wonder that few studies of physiques are attempted.

The advantage of body-typing is not quantitative, but is a means of uniquely classifying a physique by visual examination and touch (in bones, muscle, fat). Measurements of widths cannot distinguish the broad back of the endo-type from that of the meso-type, but touch and observation of muscle contour can. Body-typing gives an insight into how physiques are put together. Once you have made a dozen or so attempts at full body-typing you will find that your assessments agree with others and you can identify most types of physique 'at a glance' as one of the seven basic body-types. As a method of studying anatomy and the function of anatomy I recommend the check lists shown here.

For some work, however, broader general methods are needed, especially as a first step in constitutional research. The method which I have adopted is *body-indexing* which, while not so detailed as body-typing is totally quantitative and objective. By this means a physique can be classified in five minutes, and all you require is a tape measure. Further anatomical drawings of men and women of types 1, 2, 3 and 4 are shown in *Unnatural Selection*.[6]

Figure 3.12 Measuring the angle of the shoulders.

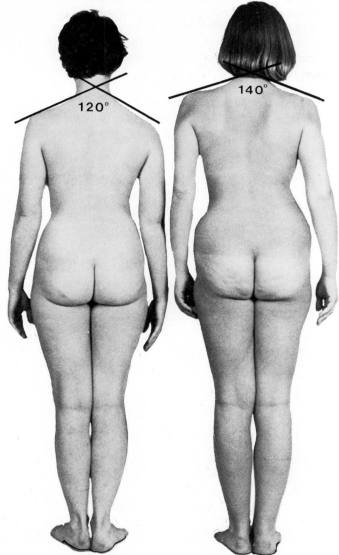

Analysis and Classification of Physique by External form

Body-indexing—a rapid quantitative classification of physique in four types

By searching the polytechnic campus for good examples of muscular physiques (meso-types), thin, fatless physiques (ecto-types), rounded types (endo-types), and examples of people who appeared to be balanced in their proportions of fat, bone, and muscle, we collected a set of mean measurements idealizing these four basic types. In all several hundred men and women in the 18–28 age range were measured. The mean values are given in Tables 3.11 and 3.12. *These are the standards for a quantitative objective classification.*

How to use Tables 3.11 and 3.12

By measuring the parameters shown in the table, and dividing by height, definitive ratios are obtained.

If a physique is homogeneous, and near to one of my standards, the ratios will fall into one of the columns, and so the body is uniquely classified, free of subjective variance. Usually a physique will score

Table 3.11 Standards for men. Ratio of physique circumference to height.

Standard body-type	1	2	3	4
Chest	.49	.55	.59	.60
Waist	.40	.42	.43	.51
Hips	.50	.53	.56	.60
Upper thigh	.29	.31	.33	.36
Ankle	.11	.13	.135	.14
Wrist	.09	.10	.11	.12
Neck	.19	.20	.22	.23
Calf	.19	.20	.21	.23
Biceps	.13	.18	.20	.21

To obtain ratios measure height, and divide into circumferences defined as follows. *Chest*: Exhaled, under armpits. *Waist*: Relaxed, exhaled. *Hips*: at pubic arch height. *Upper thigh*: at crotch. *Ankle*: thinnest point above bone. *Wrist*: thinnest point proximal to wrist bone. *Neck*: on larynx. *Calf*: standing easy, at thickest point. *Biceps*: arm hanging easily, at point midway between shoulder and elbow.

Human Measurement

Table 3.12 Standards for women. Ratio of physique circumference to height.

Standard body-type	1	2	3	4
Neck	.18	.195	.20	.21
Bust	.51	.54	.56	.60
Waist	.40	.42	.44	.49
Hips	.55	.58	.60	.64
Wrist	.09	.095	.10	.105
Ankle	.115	.125	.135	.14
Biceps	.13	.14	.16	.18
Calf	.19	.20	.22	.23
Upper thigh	.31	.33	.35	.36

To obtain ratios see Table 3.11. The bust measurement is found by measuring chest exhaled, with subject wearing brassiere. (This measurement was chosen for rapidity and convenience when measuring many subjects. It is clearly a function of thoracic growth and development of the breasts. Another standard is to measure chest as in men, and the circumference under the bust, average their values, which gives a ratio independent to breast development, but it is a cumbersome approach.)

between the column values. Thus if the chest ratio falls between type 1 and 2, it scores 2 if the ratio is nearer the value of type 2. Using this rule, any physique will score a set of numbers, which, when averaged and corrected to the nearest whole number, gives the classification of that physique uniquely and unambiguously as a 1, 2, 3 or 4.

If more categories of classification are needed the scoring is averaged to give a value to the first decimal place. If required, the standard deviation measures the mixture of the physique, and its departure from the standards. Worked examples are shown in Table 3.13. Results can be entered into a grid as shown below.

Group (e.g., athletes, businessmen, ballerinas, etc.)
Average Age
Sex

Type	1	2	3	4
Number of subjects				
%				

Table 3.13 Body-indexing of seven women.

Subject	1		2		3		4		5		6		7 (see Figure 3.9)	
Height	64.25		67.0		63.0		65.25		64.5		65.0		66.0	
Height to circumference ratio, C; score, S	C	S	C	S	C	S	C	S	C	S	C	S	C	S
Neck	0.18	1	0.18	1	0.19	2	0.19	1	0.18	1	0.19	2	0.18	1
Bust	0.51	1	0.52	1	0.54	2	0.54	2	0.53	2	0.54	2	0.53	2
Waist	0.39	1	0.38	1	0.41	2	0.39	1	0.40	1	0.39	1	0.41	2
Hips	0.57	2	0.55	1	0.56	1	0.59	2	0.59	3	0.59	2	0.57	2
Upper thigh	0.34	2	0.32	2	0.32	1	0.33	2	0.34	3	0.35	3	0.34	3
Calf	0.20	2	0.20	2	0.21	2	0.21	2	0.21	3	0.23	4	0.21	3
Ankle	0.125	2	0.119	1	0.135	3	0.130	3	0.128	2	0.138	4	0.152	4
Biceps	0.15	2	0.15	2	0.15	3	0.16	3	0.16	3	0.18	4	0.16	3
Wrist	0.090	1	0.089	1	0.095	2	0.092	1	0.089	1	0.092	2	0.101	3
Body-index		2		1		2		2		2		3		3

See appendix for full statistical analysis.

Categorization by height

By referring to Table 3.14, people can be categorized as tall (T), medium (M), and short (S). The values refer to heights without shoes.

The headings below can be used to describe groups of people.
Group (e.g., policemen, runners, dancers, etc.)
Average Age
Sex

Type	1	2	3	4
Number studied: S				
M				
T				
% each type: S				
M				
T				

Conventions used in measuring and scoring. When measuring the waist, ensure that your subject relaxes the tummy. The standards were obtained with the abdomen let out. After obtaining the ratios, score each reading by reference to the appropriate table, entering 1, 2, 3 or 4. If the ratio is exactly between two column values, enter the higher score, otherwise score to the nearest value. Wrist and ankle measurements are ratioed to the third significant decimal place, the others to the second decimal place. Once you have the scores of the various measurements, find the average of the scores, and correct to the nearest whole number.

Table 3.14 Classification of heights.

Scoring on trait lists and body index value

As discussed, any system of measurement for classification of physique gives several values for the three components of physique described in Chapter 3. The table below illustrates this, showing the spread of trait scores for the body-index. Only the rarer body-types do not fall within the values below, for example the endo-ecto-types usually have a body-index of 1, with endomorphy of four or more.

Table 3.15 Body-index scores and check list scores in 30 men (aged 18–25 years) and 120 women at PNL.

Body-index	Endomorphy	Mesomorphy	Ectomorphy
1	1	1–2	5–7
2	3–5	3–5	3–5
3	3–5	5–7	1–2
4	5–7	4–6	1–2

References
1. N.A.B.B.A. (UK Bodybuilding Association), Universe Title, 1972. Men of height between 171 and 173 cm.
2. Calculated from the data of Tanner, J. M. *Am. J. Phys. Anthrop. U.S.* 1952, **10**, 427.
3. Bullen, A. K. and Hardy, H. L., *Am. J. Phys. Anthrop.* 1946, **4** (1) p. 37.
4. Mayer, J., *Human Nutrition.* (New York: Thomas, 1972) p. 253.
5. Damon, A., *Ann. N.Y. Acad. Sci.*, 1975, **121**, 711.
6. Harris, Anthony, *Unnatural Selection?* (London: David and Charles, 1979).

Sex	Short (S)	Medium (M)	Tall (T)
Women	below { 64 in. / 5' 4" / 162.6 cm	between { 64–67 in. / 5' 4"–5' 7" / 162.6–170.2 cm	above { 67 in. / 5' 7" / 170.2 cm
Men	below { 68 in. / 5' 8" / 172.7 cm	between { 68–71 in. / 5' 8"–5' 11" / 172.7–180.3 cm	above { 71 in. / 5' 11" / 180.3 cm

4 Sexual Dimorphism

Introduction to Methods 22–26

The natural differences between male and female in our species are so obvious that anthroscopy—observation without measurement—is sufficient to identify an individual as one of the two sexes. We are justified therefore in speaking of two bodies arising from sexual differences (di (two) morphism (shape))).

At birth a male child has a penis, but the testes are not developed, and the female child has a vagina. At adolescence the differences are greater, and eventually, in maturity, the shape of male and of female differ markedly. However, we still know little about the quantitative differences between the sexes.

Morphologically, we distinguish male and female by shape (see Chapter 3 where qualitative and quantitative scales are given), while there is, effectively, an exact correlation between possession of breasts with possession of vagina, but zero correlation between possession of penis and breasts. There are rare cases of hermaphroditism.

These differences naturally reflect functional differences in sexual roles, but as a result there are differences in body size and performance. Nevertheless, although the greater muscularity of men arises through production of male sex hormones, it is not immediately obvious what advantages come from having men larger and more powerful than women. Girls grow more quickly than boys, and at 14 are often larger than boys, but are soon overtaken. Why should girls mature earlier than boys? What evolutionary purpose could it serve? Clearly the sooner the females of a species can reproduce, the greater chance of survival the species has, but this would also apply to the males. Why are more male children born than female (about 5%)?

All these differences are coded in DNA and so we have a biochemical reason for them, a mechanism to show how they occur, but functionally there are many questions we still have to answer. In this chapter are some simple techniques for quantifying differences in the male and the female. We know a lot about height, weight, and strength differences, but little about the qualitative aspects of these shape differences. Our ideas of masculine and feminine can only be improved by measurable differences, and these may help in explaining some of the problems sexual dimorphism form in evolution.

On the social level, sexual differences have moulded our social institutions and work patterns, as well as contributing to behaviour and moral codes, but often these have been exaggerated so that they no longer relate to biological realities. For example, many women are large enough and strong enough to do manual labour, and many can drive lorries, hence any embargo on them doing so is unrealistic.

Because of the importance of the subject, there is still a need for more work investigating just what women can do physically, especially considering the weight of myth, as compared with fact, still found to be believed. The evidence in the literature is suspect; what we require to remember is that women are not encouraged to physically excel. Also, we should note that the mesomorphic woman is more powerful than the ectomorphic man. The study of *type* of physique severely qualifies what follows.

Consider the following reports:

'It is commonly agreed by the medical profession and among physical educators that girls and women present certain physiological, anatomical, and emotional differences which should limit to a certain extent their participation in physical activities.'[1]

'In New York an investigation of the strength of munition workers revealed that the average industrial women had less than half the strength of the average industrial man.' 'Man is about 43 per cent muscle and women about 36 per cent muscle.'

'In proportion to weight and size the arm and shoulder muscles of the male are stronger than those of the female.'[2]

'Metabolism is less rapid in women than men, the ratio being 100 to 141. Women, biologically, are in general anabolic in type, while men are catabolic.'[3]

'The pelvis of the female is much broader after adolescence, which gives to the femur a marked obliquity. This mechanical disadvantage interferes with the running ability of the girl. In all movements of the lower extremities there is likely to be a marked lateral sway of the pelvis.'[4]

'Women and girls are not so well suited mechanically as are men and boys to participate in events requiring speed in running or unusual muscular strength.'[5]

However, we should remember that women are not encouraged to show speed and strength. I regard these findings as minimal performances in women, not what they can really do.

These physique differences are thought to be crucial:

'Differences in the type of exercise depend chiefly upon the differences in the physical structure of man and woman. Women are smaller than men and they are built in different proportions. The arms and legs are proportionally shorter, the trunk longer, the pelvis much broader. The oblique angle at which the thigh bone is attached to the pelvis interferes with the ability of women in running. Red corpuscles are less numerous in women than in men. Limited physical strength in women may be partially due to lack of use, but physiologic differences are also operative.'[6]

There remains, however, the fact that individually the ranges are enormous, and mere averages tell us nothing about the individual. A man of build 1-2-7 is very limited in strength, a 3-6-2 girl is not.

If we read *work* instead of *exercise*, we see that industrial studies should incorporate these findings and extend them for ergonomic efficiency. However, the differences are not so large as these quotations would suggest, since cultural conditioning plays a large part. Also, a mesomorphic woman can outwork an ectomorphic man. The psychological factor is very important, e.g., scores on the grip meter by women are improved if they feel it is not unfeminine to score well. It may be, therefore, that until this factor is allowed for, female performance cannot be accurately measured.[7]

Body-typing, sexual dimorphism, and Art

The following illustrations show some physical types, as studied by painters and sculptors, classified into body-types. If you go round an art gallery you can see in more detail that the interesting features are firstly that these artists, changing proportions to suit their artistic requirements, were unconsciously producing superb studies of body-types; secondly the structural differences between male and female of the same body-type are very accurately shown.

Many artists and schools show a pronounced preference for a narrow range of models. Boticelli's Venuses are usually the same girl, a rounded ecto-meso-type; Reubens used his wife and her many sisters who were all massive endo-types and endo-meso-types; while Michaelangelo liked meso-type men and women, especially in frescoes and sculptures. The Attic Greeks produced many depictions of the balanced physique. The Venus of Milo is a superbly proportioned endo-meso-type without any overweight. Renoir liked endo-meso and endo-types, while the more Gothic Memlinc liked to use ecto-types. Can you find any other preferences? (For example, look at paintings by Modigliani, Titian, and Velasquez.)

Human Measurement

Figure 4.1 (a) Rodin's young soldier has a light-boned but powerful physique. The legs in particular are heavy boned and mesomorphic, so this physique is judged as an ecto-meso-type.

(b) This physique is much more heavily boned, and more heavily muscled than the soldier. The bones make him a mesotype, since they are far too large for the ecto element to be predominant.

(c) A rounded example of the balanced physique.

(e) Rodin's plump bather has a highly endomorphic figure without a trace of mesomorphy.

(d) This girl of Cranach's reveals all the features of an ecto-endo-type, small bones yet roundedness; female contours yet no muscular relief.

(f) The smoothness of this physique with its bulk suggests endomorphy. A smaller example of the endo-meso-type.

Measuring Female and Male Shape

Method 22. Circumference method

By measuring circumferences of the body from ankle up to the neck, and measuring the height from the ground of the ankle, calf, hips, and so on, you can obtain a shape profile of the human body. In Figures 4.2 and 4.3 you can see how the waist measurement in the female falls more abruptly from the hips, in comparison to the male's, while the peak of the chest measurement in the male is higher from the hips; the reverse is true for the female. Another feature is the single front peak in the female obtained from the chest measurements, revealing, graphically, the breast, while the male's measurements fall without this feature to the neck.

These profiles show very clearly the main features of sexual dimorphism, using circumference measurements on two physiques of similar type, but differing in sex. Produce profiles for endo-types, balanced and ecto-types, of both sexes. The people whose measurements are shown in Figures 4.2 and 4.3 were regarded by the class involved in the exercise as being respectively very feminine looking and masculine looking. By collecting these profiles and comparing them, a basis for a graphical depiction of masculinity and femininity would be obtained.

Are endo-types considered more feminine than ecto-types? Or is the reverse true? You can also use the quantitative standards in Chapter 3 to construct similar graphs, and compare them with some of our measurements.

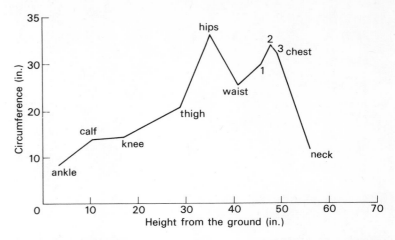

Figure 4.2 Graph of circumferences of body against the height at which it is measured of female mesotype aged 28 years. The chest measurements are: 1. below bust; 2. on nipples; 3. chest circumference above bust (see Figure 4.4).

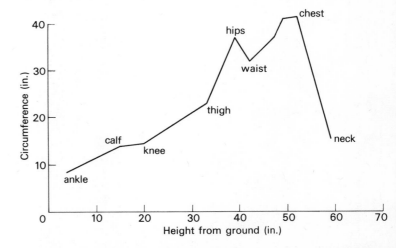

Figure 4.3 Graph of circumferences of body against height at which measured of male mesotype aged 37 years. The chest measurements are: (1) below nipple (3"); (2) on nipple, and (3) above nipple below arms.

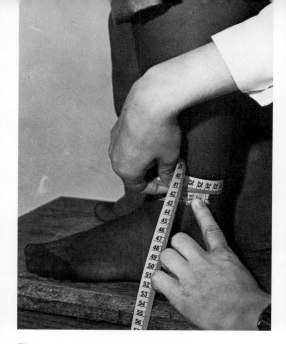 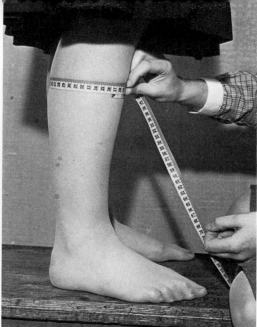 

Figure 4.4 Measurement of body circumferences from ankle to neck.

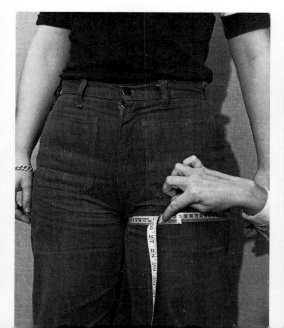  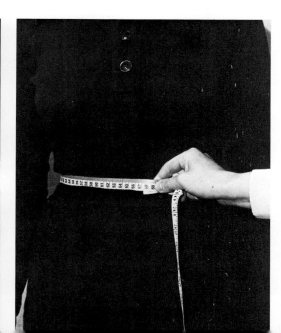

Human Measurement

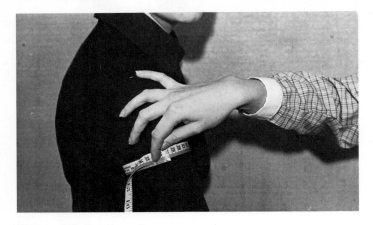
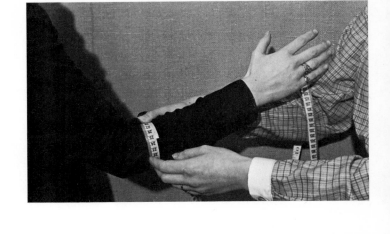

Figure 4.4 (continued).

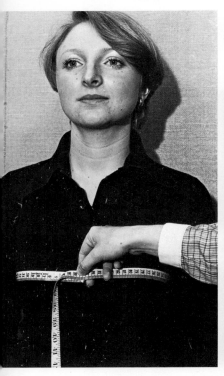

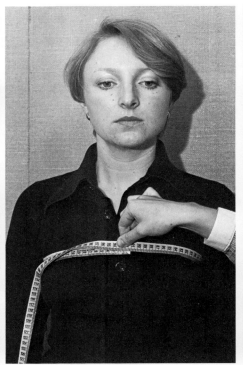
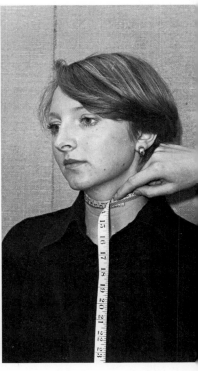

Sexual Dimorphism

Method 23. The use of width measurements of the human body to identify differences between male and female bodies

If you measure the heights of the photographs and drawings in Chapter 3 and pick various parts of the body to measure the widths (e.g., thighs, waist (see Figures 4.5 and 4.6)), you can derive a set of measurements for male and female physiques by finding the value of the ratios width/height. What conclusions do you draw regarding the relative masses of the male and female body?

In general you will find that, relative to height, men are larger in the back, while the ratio of waist to hips is lower in women than in men. By taking your own photographs you can perform the measurements at leisure on the prints.

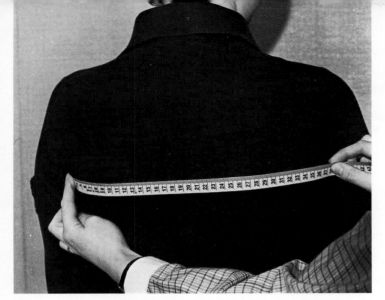

Figure 4.5 The chest measurement is taken between 3 and 4 cm above sternum and spinal column, with cavity exhaled.

Figure 4.6 The width measurement is a valid parameter, but of course is not the same as the widths shown in Figure 4.4, because *no curvature* is measured on a photograph.

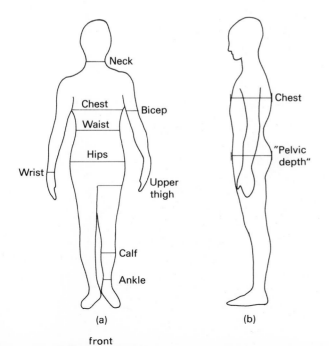

(a)　　　　　(b)

front

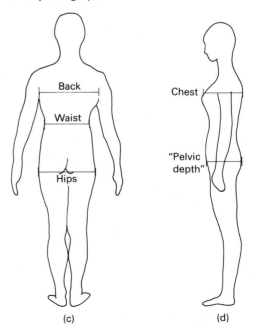

(c)　　　　　(d)

back

43

Human Measurement

Method 24. Distinguishing between male and female by measuring group properties

By taking a sample of fifty men (aged 18–26) and fifty women (aged 18–25) at the Polytechnic, and measuring height, weight, and grip strength we obtained the results shown in Table 4.1. You will observe that in all cases the means differ. In this way groups of males and females can be differentiated by simple measurements.

Perform similar tests on groups of males and females (see Part III on tests of significance of results).

Table 4.1 Measurements of height, weight, and grip strength of 50 men and 50 women at PNL.

		Height (cm)	Weight (kg)	Grip (kg)
Male	mean	177.2±6.2	63.70±6.7	53.1±10.6
	range	161–194	56.2–86.2	35.2–67.2
Female	mean	165.9±5.3	58.1±4.8	24.2±8.7
	range	153–178	44.5–76.7	14.1–41.0

Method 25. A simple method of measuring the area of the palm side of the hand

It is commonly believed that female hands are smaller than male hands, but this may be quantified. Here is a simple technique, measuring the right hand (if a person is left-handed, the left is used).

Weigh a piece of paper 100 cm² in area. Trace the outline of the hands of the people studied with a pencil on the same type of paper, and then cut out the outline (see Figure 4.7). Weigh the profiles and convert to square centimetres by the following calculation:

Weight 100 cm² = x g
Weight of profiles = y g

Therefore area of hand = $\frac{y}{x} \times 100$ cm²

The results for a small group of students at the polytechnic are given in Table 4.2.

Table 4.2 Results of Method 24 (measurement in cm²).

Subject	Female	Male
1	160.5	209.3
2	125.5	167.4
3	110.4	187.2
4	103.3	179.0
5	104.6	161.6
6	175.5	171.20
Mean	129.97	179.28

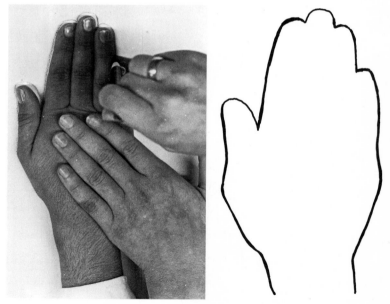

Figure 4.7 Measuring the area of a hand.

Method 26. To determine the area of the sole side of the foot

Exactly the same technique is used as with the hand (see Figure 4.8). Flat-footed men could be measured by (a) finding the foot area as with the hand, (b) taking the foot and finding the area of the profile formed. The higher the ratio of area (a)/area (b), the greater the flat footedness. What ratios do you obtain for a group of children? Is the ratio higher in girls than boys, or is there no difference?

Although female feet are smaller than males', if height is taken into account is there still a difference?

Find the flat-footed ratio of the best runners in your school and compare them with those who run less well. Is there any relationship?

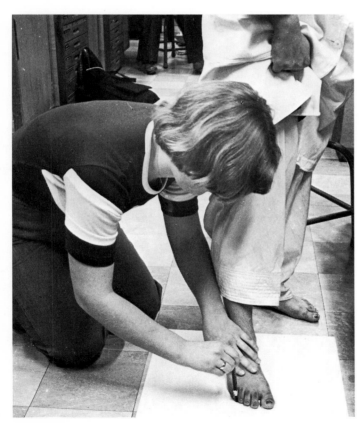

Figure 4.8 Measuring the area of a foot.

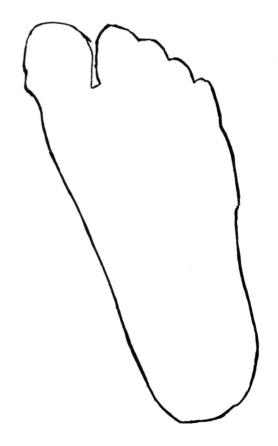

References
1. Wayman, Agnes R., *Education Through Physical Education* (Philadelphia: Lea and Febiger, 1934).
2. Williams, Dambach, and Schwendener, *Methods in Physical Education* (Philadelphia: W. B. Saunders Co., 1932).
3. Wood and Brownell, *Source Book in Health and Physical Education* (New York: Macmillan, 1925).
4. Williams, J. F., *The Principles of Physical Education* (Philadelphia: W. B. Saunders Co., 1928).
5. Sharman, J., *The Teaching of Physical Education* (New York: A. S. Barnes and Co., 1936).
6. McCurdy, J. H., *The Physiology of Exercise* (Philadelphia: Lea and Febiger, 1928).
7. An up-to-date analysis of these features is found in Harris, Anthony, *Unnatural Selection?* (London: David and Charles, 1979), while the precise range of movements needed by women is examined in Harris, Anthony, *Your Body* (London: Millington/Futura, 1979).

Part II Function

Some Methods and Measurements in Physiology, Psychology, and Nutrition

5 Weight, Surface Area, and Obesity

Introduction to Methods 27–30

As Durnin[1] said in 1967, a simple method giving fat percent in the human body would be invaluable. At present the correlation between mid-triceps skinfold and fat percent is the best basis we have in common use. From these measurements, checked against whole body weight and density measurements, it is commonly accepted that in women a skinfold between 1.5 and 2.0 cm represents a fat percentage of 15% of total body-weight, and for men, 1.0–1.5 cm represents a fat percentage of 10%. Values above these indicate fat overweight. Unfortunately, as you will have observed in your body-typing studies, fat distribution varies in different people, so what is obesity?

Fat percentage of the body weight can be found weighing people in air and in water, then using the density so obtained in the equation below (see Durnin).

$$\text{Fat} \% = \left[\frac{4.95}{\text{density}} - 4.5 \right] \times 100$$

The approximate percentage of fat in the body can be obtained from Durnin's table of percents correlated with skinfold measurements.

As previously emphasized this still does not tell us what obesity is. Curiously, from a morphological standpoint, obesity is clearly observable, but the quest is for quantitative standards. Precise body area measurements have been done by using direct estimates with squared paper and covering the whole surface. From these measurements the correlations with height and weight were found and a simple diagram can be used to give the area (see Figure 5.2). Unfortunately this method is increasingly inaccurate the fatter a subject is.

The interest in body surface area arises because the basal energy expenditure is found from it by multiplying heat loss per unit area per unit time by the area. This has importance in estimating caloric requirements. It is salutary to be aware that this most basic of physiological measurements can only be estimated to $\pm 15\%$.

In this chapter you can explore new approaches to identifying obesity and measuring the area of the human body. The aim is to reveal just how complicated these apparently simple measurements are to obtain accurately and meaningfully.

Method 27. What do we mean by obesity and how can we measure it?

Sue, Mary, and Jane were young college women, all 5′ 6″ and 21 years of age. Sue was very slim, but not angular and weighed 116 lb; Mary was of medium build and, so she and her class-mates said, was something of a college beauty at 130 lb; Jane was an excellent swimmer, broad shouldered but still, without measurements, unmistakeably female and weighed 140 lb.

If obesity is sagging flesh, none were obese. If obesity is a triceps skinfold of over 25 mm, then all were slim with measurements of at most 18.7 mm. If obesity is overweight, what is the standard? The Sun Alliance charts, which do not correct for clothes and shoes in their weighing, present frame sizes as small, medium, and large. But this subjective division is so vague that confusion is certain to occur (see Table 5.1). Alarmed at finding a range of 24 lb in the weights of three perfect physiques at the same height, I turned to the college football team. Here I found a lithe very young man of 5′ 10″ weighing 143 pounds, skinfold 14.7 mm, who was 'lightning fast' on the field, and a strapping muscular youth, also of 5′ 10″ who weighed 178 pounds with skinfold 13.9 mm. Further analyses on the campus provided youths of 5′ 10″ weighing 160 pounds with paunches, and girls of 5′ 6″ with flabby physiques at 140 pounds.

It was clear that height and weight provided no indication of obesity in themselves without anthroscopy, that is, looking at the physique. Obesity is simply identified by looking and palpating. Hopefully, we took skinfold measurements, but some people with palpably fat abdomens had scrawny arms; some women were decidedly plump-armed but scrawny in the legs. Obesity as fat distribution was clearly a complex subject. The weight of a healthy human being even at a fixed height widely varied, but deterioration of physiques through localized fat deposits is easily observed.

Fat people, even when they tense the underlying muscles, have areas on their bodies of loose thick skin, easily pinched into a handful. When the adipose layer thickens to a weight beyond the retaining elasticity of the skin, the body loses its unwrinkled contours. Pelvic bones are no longer visible, nor are the ribs of fat people. There are plump men and girls without sags, folds or creases, but the skin is tight. In other words, there is no structural deterioration.

External signs of obesity need not be implicated in men at 5′ 10″ and weighing 180 lb, or in women of 5′ 6″ and 140 lb, but some men of 5′ 10″ and 180 lb do reveal obese signs as detailed previously; and so do women at 5′ 6″ and 140 lb. Height and weight criteria are glaringly insufficient to identify obesity. It was clear that some concept was needed to identify a weight for a particular height which in a particular physique would be associated with low fat deposition.

Refusing to be drawn into the dead-end of collecting piles of statistics unrelated to a concept, an exercise which in any case had been effected many times previously by other students without their labours giving an answer to the questions I have posed above, I simply weighed and measured young women of good physique. (*Good* physique?) People who did not have sagging gluteal muscles, distended abdomens, waists with heavy fat deposits in the lumber region, or physiques with adipose blemish. My selection was highly subjective, but the measurements were objective. After several hundred examinations I could do the old fairground trick of guessing a height to half an inch, and weight within three pounds.

The subjective criteria, in addition to those already mentioned, were: narrow, medium, or broad shoulders and hips, large or fragile bones and joints; narrow or deep chest and pelvic girdle (laterally); heavily or lightly muscled. Some people had small chest volumes in relation to their abdominal size. Some people's shoulders were wider than their hips (you can tell this simply by looking at them) while others had wider hips than shoulders. Some physiques had sinuous S-shaped spinal columns, others flat. The bones of the pelvic girdle are easily felt; in some people they are large, smooth and rounded; in other people they are distinctly pointed and ridged, and much less robustly made.

Table 5.1 Heights and weights of English women in age range 18–21 and mid-triceps skinfold of less than 20 mm.

Height range (in.)	Small	Medium	Large
61			128
62	101, 102, 103, 109	110, 118, 123	123, 124
63		116, 121, 122	
64	102	107, 108, 111, 112, 113, 114, 114, 114, 126	
65		114, 115, 118, 126	
66		118, 121, 126, 127, 127, 128	132, 132, 133, 133, 140
67	119	122, 124, 125, 128, 130	153
68		133, 134, 135, 135, 135, 135, 136, 142	141, 142, 145
69	87	129, 135, 144	142
70	122	144, 146, 146, 148	

Heights are corrected to nearest inch, barefoot; weights taken nude.

Figure 5.1 Correct methods of measurement.

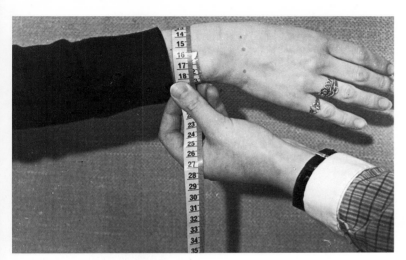

Figure 5.1(a) Measuring the wrist.

Figure 5.1(b) Measuring the ankle.

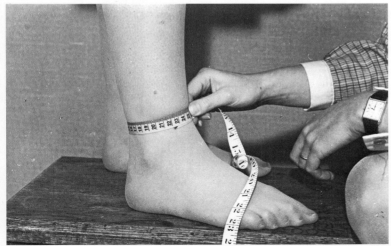

Figure 5.1(c, d) The outline of the pelvic bones are shown on a fully-clothed person marked in tailors chalk. This is not so accurate as measuring the unclothed subject.

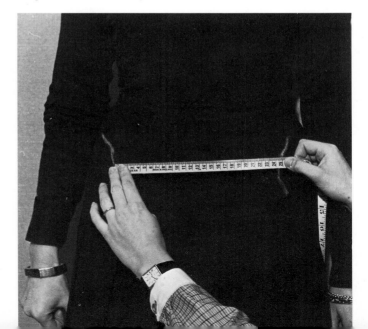

Human Measurement

Students of mine who became interested in the project were soon reaching results similar to mine, after I had shown them what physiques I was calling small, medium, and large. Constant reference to the Sun Alliance Ideal Weight tables, which I corrected for shoes and clothes, revealed that the ranges given for non-obese physiques were substantially in agreement with our own findings, but we still lacked an objective criteria for small, medium and large.

Hopefully, we set about measuring wrists, ankles and pelvic width and these measurements, along with height, were anticipated to give us a criterion of frame size. Unfortunately, we found many women with wide hips, large wrists but tiny chests who weighed very much less than some more curvaceous women with tiny ankles and wrists. Finally, after many laborious attempts at sets of measurements, I tried silhouettes. These are shown in Table 5.2. They are very simple drawings indeed, but they are objective, arrived at by trial and error over several years.

Even with the physique division into three frame types, the ranges of acceptable or ideal weight were still too large. Accordingly I used bone size measurements to divide each weight range into five. The results are shown in Table 5.3.

The weights in Table 5.3 with the accompanying figures and method of assessing the ideal or figure-weight, are very precise when compared with previous methods, and represent standards for healthy female heights of different physique types assessed by height, width, and depth, and skeletal structure.

Table 5.2 Instructions and standard for assessment of ideal or figure weight for women.

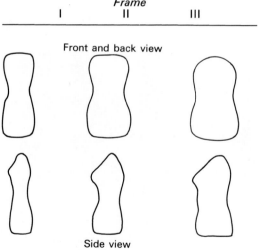

Abstractions of the female trunk from the shoulders to the top of the thighs
Frame I   II   III
Front and back view
Side view

How to measure

Wrist. Use a tape measure; the circumference at the thinnest part is used, behind the wrist bone, but not on it. Measure the hand normally used, with fingers outstretched.

Ankle. Measure the thinnest part, above the ankle bone, but not on it, standing.

Pelvic measurement. If you follow the line of the hip girdle from the waist with your thumb you will find the bones curve gradually in. Fix the point where this happens. The measurement is made directly across the stomach between the two sides of the pelvis on the top of the bone. It is essential the tape is straight and not bent out by clothing, or you will get a falsely high reading.

Bone measurements

Wrist. Less than 14.5 cm is small, between 14.5–15.5 cm medium, more than 15.5 cm large.

Ankle. Less than 20 cm is small, between 20–22 cm medium, more than 22 cm large.

Pelvic girdle. Less than 20.5 cm is small, between 20.5–25 cm medium, more than 25 cm large.

Underline the word describing your measurements, then consult the list below to find your bone category:

Category 1. Three small.
2. Two small, one medium; or two medium, one small.
3. Three medium, or small, medium, large (in any order); or two small, one large.
4. Two medium, one large; or two large, one small; or two large, one medium.
5. Three large.

Weight, Surface Area, and Obesity

Table 5.3 Ideal weights for the 5 subcategories of the three frame types.

Height (cm)	I (Smaller frames)					II (Medium frames)					III (Larger frames)				
	1	2	3	4	5	1	2	3	4	5	1	2	3	4	5
152	43.5	44.5	45.5	46.5	47.5	46	47	49	51	52	49.5	51	53.5	56	57.5
155	45	46	46.5	48.5	49.5	47	48.5	50	52	53.5	51	52.5	55	57	59
158	46	47	48	50	51	48.5	50	51.5	54	55	52	54	56	58.5	61
160	47.5	48.5	49.5	51	52	50	51	53.5	56	57	54	56	57.5	60.5	62.5
163	49	50	51	53	54	51.5	53	55	58	59.5	56	57.5	59.5	62	64.5
165	51	51.5	52.5	55	56	53.5	55	57	60	61.5	57.5	59.5	61	64	66
168	52.5	53.5	54.5	56.5	57.5	55	56.5	59	61	63	59.5	61	63	65.5	68
170	54.5	55	56	58	59.5	57	58.5	61	63	65	61	63	65.5	67.5	70.5
173	56	57	58	60	61.5	59	60.5	62.5	64.5	66.5	63	65.5	67.5	70	72.5
175	58	59	60	61.5	63.5	61	62	64.5	66	68.5	65.5	67.5	70	72	75
178	60	61	61.5	63.5	65.5	62.5	64	66	68	71	67.5	70	72	74.5	77
181	61.5	62.5	63.5	65.5	67	64.5	65.5	68	70	73	70	72	74.5	76.5	80

Table 5.4 Congruence of assessed weights with real weights of non-obese women (in pounds).

Assessor \ Assessed	Lesley		Susanna		Patricia		Angela		Chun		Deana	
Lesley 22 years	A.W. R.W.	135 132	A.W. R.W.	142 142								
Susanna 24 years			A.W. R.W.	142 142	A.W. R.W.	132 136						
Patricia 26 years					A.W. R.W.	141 136	A.W. R.W.	101 107				
Angela 26 years							A.W. R.W.	101 107	A.W. R.W.	113 109		
Chun 25 years									A.W. R.W.	113 109	A.W. R.W.	122 133
Deana 23 years	A.W. R.W.	132 133									A.W. R.W.	122 133

A.W. = assessed weights   R.W. = real weights, measured on scales.   Use conversion factor of 2.205 lb per kg.

Human Measurement

On testing any female physique the results were 90% or within 3 pounds of predicted weight when skin fold was between 13 and 20 mm, while weights outside the limits shown, when not obese, belonged to physiques with special features, such as absence of, or unusually well developed, bust, unusually long or short limbs and other such features (see Table 5.4).

Using the same inductive experimental technique construct tables for defined groups of people, stating your criteria for ideal weight. Adapt this method for males, and for younger or older age groups. Can you suggest other criteria for identifying ideal weights, and can you quantify them as successfully as the method here described for women?

It will be observed that the findings I have presented here can be put to the test in a matter of hours; all that is needed are women, a tape measure and scales. But my purpose has been more than to present a predictive method of human female weights, since the body of this essay suggests a multitude of questions. Are people with the frames and tissues characterized by the number *Frame I, height 5' 6", category 1*, different psychologically from *Frame III, height 5' 6", category 5*? Biochemically? Medically?

These experiments have not yet been done.

Method 28. Finding the surface area of the human body.

A quick measure of surface area can be obtained by measuring the circumference of the thigh between knee and crotch, t cm, finding the height, h cm, and using the relationship

$$\text{Surface area} = 2\,ht \text{ cm}^2$$
$$= \frac{2\,ht}{10^4} \text{m}^2$$

By pasting squared paper over the whole body, areas have been found and correlated with height and weight. These findings have been put in the form of a diagram relating height, weight, and surface area (see Figure 5.2).

Figure 5.2 Surface area by height and weight. To find approximate surface area, place ruler against height and weight scales above and read off surface area. (From Bell, G. H. *Experimental Physiology* (Glasgow: Smith & Son, 1947).)

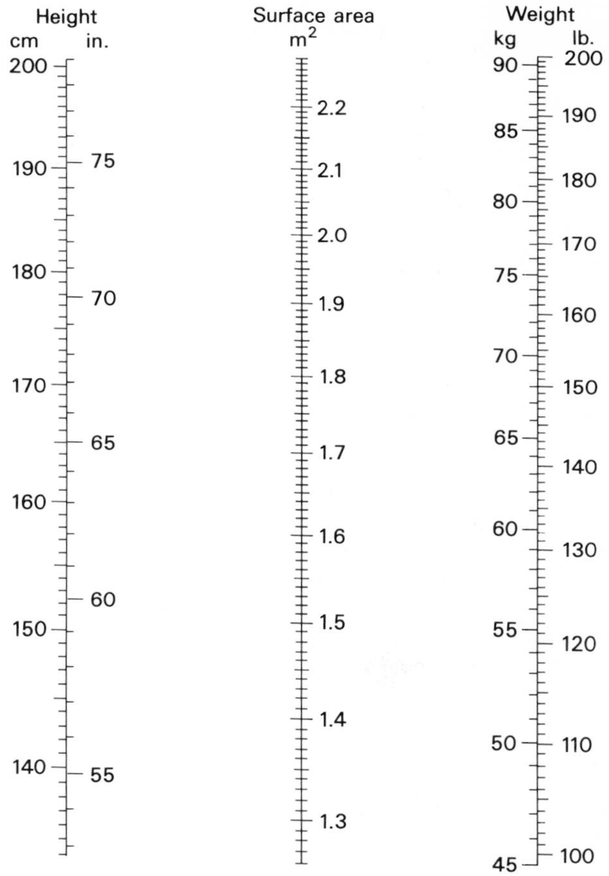

A direct comparison of the results obtained for four women are shown in Table 5.5. Compare the results you get using the two methods. Do you find that with very large physiques the discrepancy is larger?

Table 5.5 Comparison of body surface area from thigh measurement and height/weight nomogram for four women.

Subject	Surface area (m²)	
	Thigh measurement	Height/weight nomogram
1	1.41	1.40
2	1.52	1.51
3	1.79	1.71
4	1.67	1.65

Compare your results with areas found using the following technique:

If you imagine a thin segment of, for example, the leg, of thickness dT and circumference C, the cylindrical area of the segment is CdT. To get a very accurate value of the area of the leg you would need to keep dT very small and sum over the whole leg. Mathematically speaking, the area of the leg is

$$\int_{Ta}^{Tb} CdT$$

and if you knew T = function of C you could find the value of the integral between Ta and Tb, where a is the lower limit and b the upper limit of the leg. A graph of T, the height of a point on the leg from the ground, versus C, the circumference at that point, will yield a curve. The area under this curve is the area of the leg between Ta and Tb. The more values of T you find between Ta and Tb the more accurate will be the result.

You can try this if you wish, but a simpler method is to note that the integral is approximately equal to the sum of measurable segments of the leg.

$$\int_{Ta}^{Tb} CdT \simeq \Sigma C \Delta T \simeq \text{average circumference} \times \text{length of leg } (l)$$

Refer to Figure 5.3 where various circumferences are identified as C_1, C_2, C_3 etc. These can be used to find the area of the leg as follows.

Area of leg

Find $(C_1+C_2+C_3+C_4+C_5)/5$ and multiply by l, giving the area of leg, A_{leg} (see Figure 5.4). Total area of legs = $2A_{\text{leg}}$.

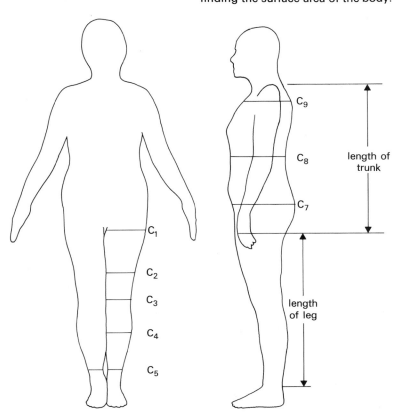

Figure 5.3 The circumferences.

Figure 5.4 Measurements used in finding the surface area of the body.

Human Measurement

Use the same technique to find the area of trunk, using C_7, C_8, C_9 (or more points if you wish greater accuracy). Proceed similarly for the arms and hands. The head area is given approximately by taking two circumferences, averaging them, (one horizontal, across the forehead, one under the chin as in Figure 5.5) and using the formulae for the area of a sphere substituting

$$r = \frac{\text{average circumference}}{2\pi}$$

Sum the totals over the whole body to get the total area of the body. The degree of accuracy of your result will depend on the number of measurements you take, and the greater number of segments you divide the body into.

Since we have so few results in this subject, valuable results could be obtained by picking boys, or girls, of the same height but different weights, and tabulating area versus weight values. Once one height had been worked out, another height could be investigated.

Lean body-weight

This is the weight of the body minus fat. If we assume density of fat is unity (see Durnin), and the skinfold, averaged over the points shown in Figure 5.6, then the weight of the fat layer is

$$\frac{\text{Area of body} \times \text{skinfold}}{2}$$

The 2 appears because we assume that folding the skin doubles its thickness.

This analysis assumes that the skinfold is indeed twice the fat layer, and can only be a very rough approximation, as is the assumption of the fat density as unity, but the method gives a first order of the fat weight of the skin and the lean body weight. The calculations are shown below for three young women who appeared to have excellent weights. Average fat thickness, t cm, for three young women over points shown in Figure 5.6 were

(1) $t = 1.56$ (2) $t = 1.45$ (3) $t = 1.43$

These values were obtained by dividing the 'pinch' thicknesses by 2.

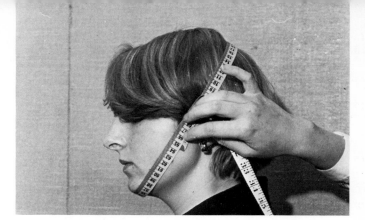

Figure 5.5 Measuring the circumference of the head.

Figure 5.6 Points where the skinfold should be measured to find lean body-weight.

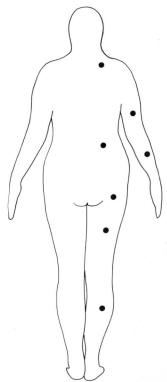

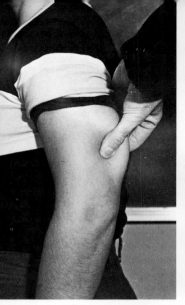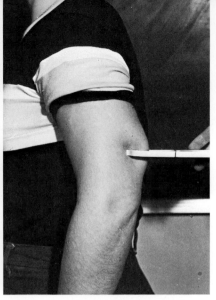

Figure 5.7(a, b) A simple method for assessing thickness of adipose layer on the triceps as shown here. Make sure you are not gripping muscle by asking subject to lock the elbow: fat does not get harder, muscle does. Cheap calipers can be used (see apparatus appendix).

If the weight on the scales is W kg, the area of the skin A m² then the weight of fat is approximately At kg, and the lean body weight $W - At$ kg. The % fat is

$$\frac{At}{W} \times 100.$$

Using the circumference method of finding A, the % fat of the subjects were 1 : 22.3; 2 : 21.2; and 3 : 20.6 which agreed with the % fat using the Durnin caliper technique.

Method 29. How weight changes with circumference changes

If the circumference of a circular cross section is c, and the length is l, with density p, the weight of this cylinder w is given by

Weight, Surface Area, and Obesity

$$w = \frac{c^2}{4\pi} p, \text{ or } w = kc^2, \text{ where } k \text{ is a constant.}$$

This means, on differentiating, that,

$$\frac{dw}{dc} = 2kc, \text{ or } dw = 2\, kc.dc$$

Since $w = kc^2$, we have

$$\frac{dw}{w} = \frac{2dc}{c}$$

In general this relationship holds for small real changes Δc of the circumference giving the weight change as Δw,

i.e. $\quad \dfrac{\Delta w}{w} \simeq \dfrac{2\Delta c}{c}$

If we make the rather large assumption that the cross section of most parts of the body are roughly circular, and the density is roughly uniform, we should be able to predict the approximate weight changes occurring when, for fixed height, the circumference changes.

A worked example is given below, where you will see the results are promising, considering the simplicity of this analysis.

Further exercise

(i) Using this analysis, predict the relative weights of ecto, balanced, meso, and endo-types from the circumferences in the tables, of Method 20, Chapter 3.
(ii) Weigh two persons of equal height and calculate their average circumferences. If w_1 and w_2 are their weights, and c_1 and c_2 their average circumferences, then

$$\frac{w_2 - w_1}{w_1} \simeq 2\frac{(c_2 - c_1)}{c_1}$$

Is this true for your pair?

(iii) In general the contribution of the fat layer to the average circumference in non-obese people is small, but it is larger for healthy girls of 14+ than for healthy boys of 14+. Often people think they are obese when they are not, especially girls. A study of body-types helps put this matter in perspective, but a very simple test is this: pair off people with different builds, one of whom is slightly built, but who are approximately the same height. See if the relationship in (ii) is approximately true. (It will hold less well when the value $(c_2-c_1)/c_1$ is greater than 10%.) If the relationship holds, and the fat layers of both people are comparable (use calipers) at different parts of the body (for example, mid-triceps, tummy, side of waists, inside thigh), then the weight difference is one of frame size, not fat.

Worked Examples

Percentage weight change

$$= \frac{\text{twice average circumference change} \times 100}{\text{average circumference}}$$

Table 5.6 Measurements of a 30-year old male subject. The values A, B, and C were obtained at different times over a period of 2 years.

	A	B	C
Weight (kg)	69.40	73.90	79.8
Circumference (cm) of			
Chest	104.10	104.10	106.70
Waist	78.70	82.60	86.40
Hips	94.70	95.25	101.60
Upper thigh	52.00	55.20	58.40
Biceps	30.80	32.40	33.00
Wrist	17.80	17.80	18.10
Ankle	21.60	21.60	21.90
Average circumference	57.00	58.42	60.87

(A) Percentage weight change

$$= \frac{2 \times (58.42 - 57)}{57} \times 100$$

$$= \frac{2 \times 1.42 \times 100}{57}$$

$$= 4.9890$$

∴ new weight

$$= \text{original weight} + \frac{(\text{original wt} \times \% \text{ wt change})}{100}$$

$$= 69.4 + \frac{(69.4 \times 4.98)}{100} \text{ kg}$$

$$= 69.4 + 3.46 \text{ kg}$$

$$= 72.85 \text{ kg}$$

(B) Percentage weight change $= \dfrac{2 \times (60.87 - 57)}{57} \times 100$

$$= \frac{2 \times 3.87}{57} \times 100$$

$$= 13.6\%$$

∴ new weight $= 69.4 + \dfrac{(69.4 \times 13.6)}{100}$ kg

$$= 69.4 + 9.43 \text{ kg}$$

$$= 78.83 \text{ kg}$$

A more accurate approach is to calculate the percentage changes of each circumference, e.g., a man had the measurements given in Table 5.7.

The average percentage increase of circumference is 7.05%. Therefore increase is percentage weight $= 2 \times 7.05 = 14.1$. Original weight $= 155$ lb.

$$14.1\% \text{ of original weight} = 155 \times \frac{14.1}{100} = 21.86 \text{ lb.}$$

∴ predicted weight $= 155 + 21.86$

$$= 176.86$$

$$= 177 \text{ to the nearest pound.}$$

Therefore the predicted weight and actual weight differ by only one pound. Note, too, that the percentage change in circumference of wrist and ankle is usually very much smaller than the other circumferences, and in this calculation they have been omitted. Use this method or the previous examples and check if the results have improved.

Table 5.7 Two sets of measurements for a male subject, with the percentage increase.

Weight	155 lb	176 lb	Increase	Percentage increase
	Circumferences (in.)			
Chest	40.5	42	1.5	3.70
Waist	32	34	2	6.25
Hips	37	40	3	8.11
Biceps	12	13	1	8.23
Thigh	20.5	23	2.5	12.20
Calf	13.5	14	0.5	3.70

Method 30. Assessment of obesity by comparison with standards

Determine the ratios of the circumferences listed in Table 3.7 or Table 3.9 for the person you are studying. Obesity is immediately recognised by these measurements alone if the chest (or bust) measurement is exceeded by the waist measurement. In the case of ectotypes, however, this event can occur even with no trace of fat, since the abdominal muscles are often so slack that the shape is distorted. In pregnancy, in all body-types, the value of bust/waist is less than unity.

Incipient overweight is also shown when the values waist/height, hips/height, chest (bust)/height are all similar, but if the chest/height ratio is similar for the ecto-type, these findings are not indicative of overweight. Since obesity implies an excess of fat, the measurements of the body must depart from the standard values for the different body types given in tables.

When using these criteria it must be remembered that some types of obesity are localized in hips and thighs. Since some perfectly healthy and ideally weighted physiques have measurements pertaining to endo-types in the hips and thighs, but balanced types, or even ecto-meso-types in the upper and middle part of their bodies, ratios alone cannot identify obesity in such physiques.

An upper range to the ideal weight can be calculated for any obese physique in the following way:
(i) Determine the circumference ratios to height as noted in Tables 3.7 or 3.9.
(ii) Identify the body-type from these ratios by comparing them with the standards, leaving out the waist ratio.
(iii) Calculate from the PI of this body-type x (see Table 3.6) and the height of the individual h, the standard weight w_s

$$w_s = \frac{h^3}{x^3}$$

(iv) w_s is the approximate upper limit to the ideal weight for that body-type. Weights above are usually obese weights.

Exercises

The standards given in the book generally refer to adults between 18–25. Body-types using the qualitative lists and methods given in Chapter 3 are applicable to girls from 14 on, and to boys from 16 on, but the drawings refer to the 18–25 age range. Attempt to construct your own standards for the younger age range. Also try to classify a group of younger children. If you are lucky enough to be in a large school, then you could follow individual children, quantitatively and qualitatively, hopefully with photographs as well, from 11 to 19. This has never been done in the detail described in this text. *Providing you clearly identify your methods*, such work could prove of great value in conjunction with other results, when collected, and properly analysed.

References
1. Durnin, J. V. G. A. and Rahamau, M. M., *Brit. J. Nutrition*, 1967 **21**, 681.

6 Physiological performance

Introduction to Methods 32–37

Physiological performance of your body can be measured by simple methods and also by very complicated ones using elaborate apparatus, but the aim is always the same, to find out how your body behaves when resting and how different activities alter this resting value. We know that very fit athletes can push up their pulse rate to nearly 200 beats per minute, but within a few minutes after the exercise it is back down to normal, about 70 counts per minute. Less fit people cannot do this, and in general the fitter you are the more quickly your body returns to normal after a stimulus, whether it is physical, chemical, or emotional.

What we have learned to do is use, as far as possible, a single measurement which tells us a great deal. The most useful one is counting the pulse rate. People whose heart rate falls back quickly after exercise also show healthy patterns of electrical behaviour of their hearts, indicating good heart muscles. In other words the pattern of heart rate changes tells us a great deal, even suggesting to us that when the pattern is good, lung function is good too, getting oxygen quickly to the tissues and removing carbon dioxide and lactic acid from the blood.

When you are frightened your heart rate rises, because hormones, chemically called catecholamines, are released into your blood stream. Healthy people's hearts soon revert to normal, so the pulse rate tells us about mental and emotional stress too. Very calm states of mind are related to very low pulse rates, even as little as fifty counts per minute, a characteristic of many great runners. The pulse occurs because the heart contracts forcing blood through the arteries, so any artery can be touched to feel this effect. How many pulse points can you find?

Method 32. Measurement of pulse

You can feel blood being rhythmically pumped through arteries at many points in the body, but convenient ones are on the wrist, palm side; at the base of the thumb (see Figure 6.1); and at the side of the neck near the windpipe, just above the larnyx. Touch these parts on yourself and others, using the tips of your first and second fingers.

Can you identify other pulse points? If so, you could trace the main arteries in the body. Are the pulses exactly the same beat on left and right wrists?

You can determine the pulse rate simply by counting the pulses performed by the heart in a minute. A more rapid but much less accurate way is to count the number in 15 seconds, and multiply the result by four, but this can give an error of as many as 10 counts per minute. Because of the importance of pulse rate and because it is so difficult to count rates above 130 pulses per minute, in athletic and drug research the pulse is recorded automatically by electrical monitors, either at a pulse point or by measuring the associated electrical changes in heart muscle. Nevertheless, our knowledge was based on simple manual counting and excellent results can be achieved. Count the pulse rates of someone sitting down, and standing up. Is there a difference?

Measure the pulse rates of the class with people lying down for at least five minutes. You will see how the values vary from person to person. Find the average for a group of boys and for a group of girls, plot the results on a scatter diagram (Figure 8.2). Usually the highest and lowest values for boys are more spread apart than for girls, and the average is higher. Can you see that with your results? Can you see the bunching of girls' values? This is a common feature of many measurements; women's values are less spread out, more close together, less extreme.

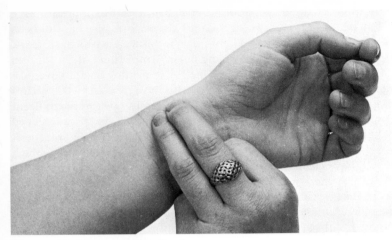

Figure 6.1 Measurement of pulse.

Method 33(a). The pulse rate as a measure of work

To do work we burn food, producing carbon dioxide, water and phosphate-containing molecules. To fuel this fire, oxygen has to reach our tissues, so the heart increases the rate of its work. It should then follow that a higher pulse rate shows a greater amount of work. Is this true, and in what way is it true? The way to test this hypothesis is to see if it works for individuals, in other words, to see if your heart rate goes up the more work you do (or you can use a friend as a guinea pig).

Get a strong box, or the gym benches, and step on and off it, wearing gym shoes or no shoes or socks. If you do it fast your pulse goes up. You are working against gravity as you stand up onto the box.

Try the following measurements. Step on and off the box ten times in 45 seconds. Measure your pulse. After your heart has returned to its resting value (see Method 32) do the same amount of work in 30 seconds, that is, step on and off ten times in 30 seconds. Measure your pulse.

Repeat the work, with resting intervals, over 20 seconds, 15 seconds, and 10 seconds, and then finally for the smallest time you can do 10 steps in. Plot a graph of your pulse rate against time to do the ten steps. Can you see that the same work done in a smaller time makes your heart work harder? The explanation is very simple—can you give it?

If you compare your graph with someone else's you will see that the curve is different. Could it be that for the same type of work different people respond differently? Is it fair to compare the results of a big person with a smaller person in this way, because the bigger person does more work, doesn't he?

Method 32(b). Heart recovery

If your heart is again beating at its resting level a short time after exertion this suggests that it responds very efficiently to your needs, beating fast when you need oxygen quickly, but resting itself when the job is done. We know that athletes' hearts fall back to a steady rhythm very quickly after even the most strenuous competition, sometimes back to normal within two or three minutes.

How quickly does your heart return to normal after doing 30 steps on the box in 30 seconds? To do this, measure pulse rate at the beginning, then after the exercise count pulse rate every minute for as many minutes as it takes to reach normal again. Collect other people's values. Do you see if physique has any effect on the results?

Method 34. Limits in physical performance

The distances achieved in Olympic javelin, discus, shot, long jump, and high jump competitions have all steadily increased. Clearly, there are limits to what ultimately may be achieved, but can one assess where these limits lie? The 100 metres race hardly requires a breath during the event, it being run on the oxygen and energy supplies in blood and muscle: long distance races need vast consumption of oxygen, and there was less oxygen per lungful at Mexico than at the other venues. The 100 metres, therefore, is essentially a dash against gravity, and wind resistance, the oxygen factor being less important. It might reasonably be argued that the speed reached in the 100 metres is the ultimate that a man can expect to reach, that it is an all-out effort. Can a man keep the maximum effort up for 200 metres?

In the 1968 Olympics, this was realised with times of 9.9 seconds for the 100 metres, and 19.8 (2×9.9) seconds for 200 metres (1.0101 metres per second). There are grounds for using the times for 200 metres rather than the 100 metres as the ultimate race because some top class athletes, perhaps the majority, are still accelerating at the tape. Jesse Owens could be seen on Riefenstahl's film to be accelerating up to the tape. Possibly one day someone will do the 200 metres faster per metre than the 100 metre race. The 400 metres should be run, in mathematical terms only, in 2×19.8 (or 4×9.9) seconds, that is 39.6 seconds, but the fastest recognised time at the time of writing is 43.8 seconds for the Olympic Gold, and indeed the world record is this time as well.

We have, as an 'ultimate' target for running,

$$\text{Record time (s)} = \frac{\text{distance (m)}}{1.0101}$$

In the 1968 Olympics, Tommie Smith in the 200 metres, and James Hines in the 100 metres, both of the USA, required only 0.99 seconds to cover a metre of ground. Lee Evans required 43.8 seconds to cover 400 metres, namely 4.2 seconds more than the value of 39.6 seconds predicted by the formula. This figure can be regarded, safely I would suggest, as the ultimate time for the 400 metres (until of course some new super-athlete proves mathematics wrong).

The fact is that, on present form, no one is running 800 metres in only twice the time it takes to do the 400 metres in Olympic events. In fact, the time taken to cover a metre gets progressively longer as the length of the event increases; it works out at 0.143 for the 1500 metres, 0.1304 for the 800 metres, and 0.1095 in the 400 metres, the figures in each case representing seconds per metre, using data from the 1968 records (compare the latest world records).

In 1956, Vladimir Kutz ran the 5000 metres in Melbourne, taking 0.164 seconds on average per metre. That performance has not been bettered in any Olympic Games at the time of writing. In 1964, William Mills of America ran the best Olympic time of 28 minutes 24.4 seconds for the ten thousand metres, indicating that he covered a metre on average every 0.170 seconds. Bikila in fact ran 10 000 metres at a rate of 0.173 seconds per metre. Those who saw this magnificent runner enter the stadium will know that he had not reached exhaustion, he was running smoothly and strongly. The figures speak for themselves; he had run at a rate over thousands of metres, which would have given him excellent performances in smaller distances.

As a runner progresses, the amount of lactic acid in his blood increases measurably as a result of oxygen shortage. However, considerable amounts of energy can be derived from the sugars in the blood and muscles without the intervention of oxygen, by a process well known in biochemistry, termed glycolysis. Oxygen is required to keep the oxidative processes continuing, and is directly concerned with keeping the level of lactic acid in the blood low. In the 100 metres, little oxygen is used to gain energy, and it is produced by the glycolytic mechanism and the oxygen already in the blood. It is during the prolonged races that the athlete must have *oxygen* during the event. Training may have an improving effect on the rate at which oxygen reaches the tissues from the blood, but unless the oxygen is taken into the lungs quickly during the event, the muscles cannot function long. Since bravery is conspicuous in athletes, the respiratory function may well be the limiting factor rather than pain. Thus although we can expect records to continue to be broken, the rate at which they are broken must surely at some stage decline.

Figure 6.2 shows the steady improvement in the Olympic 1500 metres from 1900 to 1968. At present, this rate of improvement would imply a record time of 2 minutes 58 seconds in the 2024 AD Olympic Games. As our knowledge improves, will athletes be able to toughen their muscles, improve their blood function, and through guts and training improve their respiratory function, *and* on the day of the Olympiad, produce this consummate performance? It is very unlikely, but how close will someone get?

Physiological performance

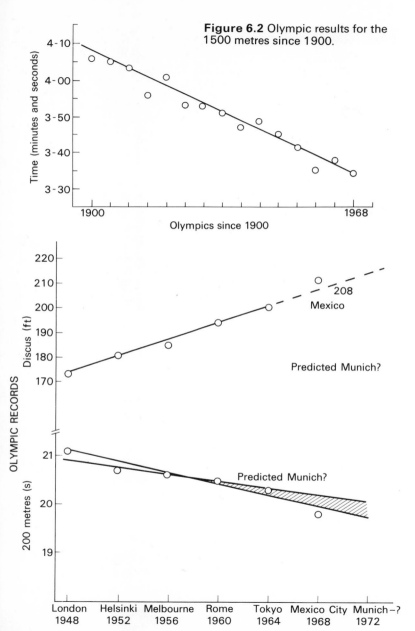

Figure 6.2 Olympic results for the 1500 metres since 1900.

Figure 6.3 Predictions in discus and 200 metres on past Olympic records.

Exercises

1. Calculate regression equations from Olympic records up to 1968, to 'predict' '72, '76 and 1980 achievements (see appendix for methods of calculating regression equations).
2. Redraw Figures 6.2 and 6.3 to include present day data.
3. Investigate your school or college records for similar trends of improvement.
4. In Figures 6.2 and 6.3, I have simply plotted the Olympic records versus year of event. Observe how the improvement follows a clear linear trend. The lines drawn are 'best guess' lines, by eye alone. Using the same data, calculate the regression equations as described in the appendix. Are your predictions more accurate using statistical methods than merely extrapolating 'best guess' lines?

Method 35. What is strength, and is it related to pulse rate and recovery?

Strength is a word we all use, but it is very difficult to measure, because we mean so much by the word. Is strength the ability to lift heavy weights? Yes, it is, but what about the difference in strength of a man weighing only 150 pounds who lifts 300 pounds, and a very large man of 250 pounds who can lift 400 pounds? The latter is stronger in the sense that he lifts more, but weaker in the sense that he cannot lift as much, pound for pound of his body weight. And what of endurance? A woman who can run a 1000 metre race at Olympic standard has a different kind of strength from weightlifters, and doesn't a marathon runner have to be strong? What features would you suggest need to be included in a definition of strength?

However, it is interesting is to see if strength and pulse rates are related in any way, but first we need a definition of strength. In the following experiment we tried to see if the ability to do a lot of work quickly was in any way related to grip strength (Method 13). This is what we found.

Ten women aged 19–22 each lifted a fourteen pound weight with

Human Measurement

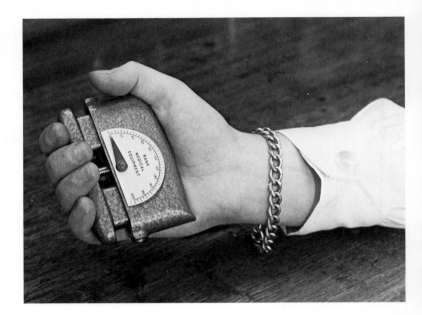

Figure 6.4 Method of measuring grip strength.

Table 6.1 Sample results showing values of strength and grip.

Subject	Number of times weight of 14 lb lifted in 30 s	Reading from dial on grip meter
1	8	16
2	11	20
3	12	18
4	15	25
5	14	22
6	6	15
7	12	20
8	11	19
9	12	20
10	13	21

their writing arm from waist to shoulder height by bending the arm as much or little as they pleased, but as many times as possible in 30 seconds. After resting each person had their grip measured using a Rank grip meter (three readings, average taken).

A graph, Figure 6.5, was drawn and you can see how very well correlated the two values are. In other words the better your grip the better your ability to do this work. Note however, you cannot really draw a line for the extreme points because as in all human measurements the correlation is not exact and we cannot tell on this evidence if the line does bend or if it really is straight. However there is sufficient data to suggest that one can bet fairly heavily on a strong grip person as one with 'strength'.

Try this experiment yourself, picking a suitable weight for your age group, measure pulse and recovery after the work is done. Plot these values against grip strength *and* ability to do muscular work. How good is the correlation? Do you think heart recovery measurements

Figure 6.5 Graph of strength versus grip. Strength was measured as performance in the repetitive lifting of a weight and grip measured by a spring-loaded metre with arbitrary scale.

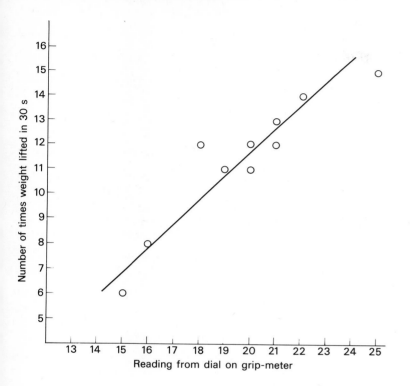

are essential for a realistic assessment of strength? What other factors would you like to see included?

Method 36. Body-weight and strength

If we plot the total lifts for the 1972 world records against the weight class of the athlete, as expected the total performance in the three lifts (press-squat-jerk) increases with body-weight. However, as you can see from the graph (Figure 6.6) the exertion or strength falls off from linearity at 65 kg. In other words after 65 kg the human frame can exert less, kilogram lifted per kilogram body-weight, so that with the very great body-weights (super heavyweights) the weight lifted is some 25 kg less than would be expected from the performances of those below 65 kg in body-weight. Can you explain this? For example, muscle strength increases with its dimensions as a square, but muscle weight increases as a cube of its dimensions, indicating that at some weight the muscle will be so heavy it cannot contract against its own weight. Investigate other efficiency versus body-weight relationships —for example grip strength versus body-weight.

Figure 6.6 Body weight of lifters versus weight lifted (1972 World records).

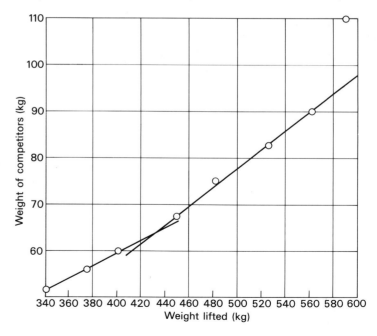

Human Measurement

Method 37. Modified Harvard Fitness Test

You will need a stopwatch and a sturdy stepping platform 20 inches high. People doing the test should wear gym gear and plimsoles.

At 'start', place one foot on the platform, step up placing both feet on the platform, straighten legs and back. Immediately step down. Bring down the same foot first that you put up first. Go up at exactly 2 second intervals. It helps if the time count is called out by an observer. Count time when the subject starts, exercise at the rate of 30 times per minute for 5 minutes. At 'stop', sit down, count time. Begin pulse count exactly one minute after stop. Count the number of heartbeats for exactly 30 seconds.

Record the number of heartbeats in the 30 second period one minute after the exercise. Inspect Table 6.2 for the numerical score, i.e., if pulse count is 50, the score is 100.

Healthy young men's scores are graded as:

Poor: less than 50   Average: 50–80   Good: above 80

Women can use the same scale, but scores are adjusted 10%, e.g., a woman's score of 50 is equivalent to $50 + 10\% = 55$.

Table 6.2 Scoring in Modified Harvard Step test.

	Pulse count (in 30 s)					
	40–49	50–59	60–69	70–79	80–89	90 and over
Score for 5 minutes effort	120	100	85	73	65	60

References

For examples of papers dealing with advanced work in this subject see:

1. Brooke, J. D. and Hamley, E. J. (Human Performance Laboratory, Physical Education Section, University of Salford and Department of Ergonomics, University of Technology, Loughborough, England), The heart-rate–physical work curve analysis for the prediction of exhausting work ability, *Medicine and Science in Sports*, 1972, vol. 4, no. 1, pp. 23–6.
2. Brooke, J. D., Hamley, E. J., and Thomason, H., Variability in the measurement of exercise heart rate, *The Journal of Sports Medicine and Physical Fitness*, 1970, vol. 10, no. 1, pp. 21–6.
3. See Khosla, T., *Lancet*, 5 Jan. 1974, p. 30 and Ryder, H. W., Carr, H. J., and Herget, P., *Scientific American*, June 1976, p. 109, for approaches to the problems suggested in this chapter.

7 Homeostatic mechanisms

Introduction to Methods 38–40

Claude Bernard gave us the concept of a fixed or steady resting state in our physiologies, so that we can expect a steady pulse and fixed levels of materials in our blood. We know that these levels vary when we are stimulated. After a meal, glucose in the blood rises from the usual 80–100 mg per 100 ml to around 160 mg %. Above this level, glucose appears in the urine. Usually the level returns to resting or normal levels because glucose is oxidized while the rest is converted to glycogen in the liver. This picture of a departure from normal levels to a peak, because of a stimulus, and the falling back to normal levels, is a common feature of our physiology and biochemistry. The tendency we have for fixed levels is termed *homeostasis*, and the multitude of mechanisms we have to achieve homeostasis are called *homeostatic mechanisms*, many of which are hormonally controlled.

Stress can be thought of as the result of saturating the ability our bodies have for returning to normal levels, while we now know that our resting state is not fixed as Bernard supposed, but alters periodically. Here are some new exercises which demonstrate these features of homeostasis.

Method 38. Analysis of response to stimuli.

In considering the human body and its response to various stimuli, it has long been recognized that there exists levels of stimulus which create disturbances in the homeostatic mechanism which are not easily reversed. In extreme cases, injury or death results. Furthermore, we are also accustomed to think of threshold values of stimulus, below which a discernable change is not shown.

If we call a stimulus S, and a response R, we know that for numerous R and S pairs the following conditions apply: that the equilibrium value, R_0 is undisturbed for all S less than S_0 ($S<S_0$), where S_0 is the threshold value; that for S greater than or equal to S_0 ($S \geqslant S_0$), R varies from R_0; and that for some value of S, $S \geqslant S_{max}$, R either stays constant, or varies in an entirely unpredictable manner (see Figure 7.1).

Consider the oral administration of glucose. This is the stimulus S, and the response can be measured by the glucose level in the blood. For normal subjects the resulting curve of R versus time is typically 'sinusoidal', i.e., the level rises with time, then falls. For diabetics, the curve flattens out. If the dose is increased, death eventually intervenes in the diabetic, while the normal subject shows a decreasing ability to return to R_0.

Figure 7.1 Response–stimulus time curve.

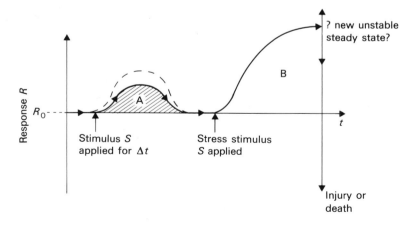

When considering $S-R$ relationships, it must be realised that much work in the literature is of the 'pulse' type, when S is applied for a short time, Δt. Much other work is concerned with monitoring R versus S with S continuously applied at a fixed level.

'Pulse' effects (see Figure 7.1)

An obese man runs, his heart rate goes higher and higher—the rising curve of R. He stops running (S removed after Δt), his pulse returns to normal (area A). If he runs faster, a higher R_{max} is found. However, suppose he runs so hard that he has a heart attack, with arteries ruptured. There is no return to normal levels, the smooth curve of pulse rate falls away to zero *discontinuously* (area B, R falls to zero).

A muscle will contract when stimulated with an electric current; as the rate of stimulus increases, the rate of contraction increases, but at a certain frequency the *muscle will stay contorted. The rate of contraction has fallen to zero* (area B).

A tendon stretches during wrestling (a hold held for Δt). The hold is released, the tendon returns to normal. But if the hold is applied with sufficient force the tendon ruptures (R rises). For this case, R has, discontinuously, risen sharply.

These are 'pulse' effects of a fixed S applied for Δt and R is examined *after* the application of the stimulus for a longer time than Δt (see Figure 7.1). For other examples see Adolph.[1] The relationship between S and R can also be studied when S is continuously applied at increasing intensity over a time period (in effect the maximum, or peak value, is the one recorded).

Examples of continuous application (Figure 7.2)

Curves a, b. Elasticity (R) of arteries in middle-aged (a) and young people (b) against applied pressure (S).

Curve c. Energy expenditure (R) as a function of walking speeds. Modified from Durnin and Passmore.[3]

Curve d. Pressure of O_2 in lungs (R) with increasing altitude (S). At about 25 000 feet the curve suddenly dips because O_2 will not stay in the blood.[4]

Table 7.1 Some common measured stimuli and responses.

S	R
1. Severity of Exercise	blood lactic acid levels and pulse rates
2. Electric Shock	muscle contraction
3. Temperature Changes	pulse rates

Figure 7.2 Response and stimulus graphs. Response (R) and stimulus (S) in arbitrary units drawn arbitrarily on axis to show different shapes of curves. These curves are plots of R and S, but the sequence of events for applying S is for the response to reach a maximum, recorded here, when S is removed R returns to original value.

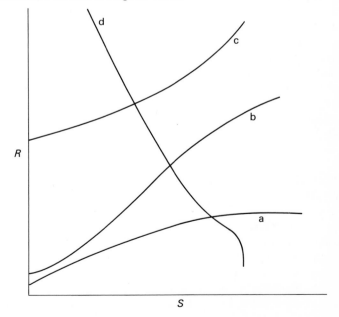

When the stimulus is reduced to zero and the minimum response reached, the system will return to its resting state. This is a dampening effect, which brings the system back to R_0. Dampened oscillations are described by the function $R \propto \sin wt/e^{kt}$, where t is the time after the stimulus is applied, k is a constant, and w is measured in radians per second. These values can be calculated for simple mechanical $S-R$ relationships, but for physiologic functions they have to be constructed by experiment (very few have been). The damping increases with increasing values of k. In man, k is usually large enough to make the value of R effectively R_0 after one period ($t = \pi/w$), but it has been shown that voluntary muscle contraction after a single stimulus describes a perfect damped oscillation for several periods.[5]

We know too that changes in R increase in magnitude as the magnitude of S increases, but providing the sine effect is operable, and the value of k sufficiently large, the organism recovers and returns to R_0. For very efficient systems R returns to R_0 quickly, which means w is large. Providing that R varies with $\sin wt/e^{kt}$, even with large S values, no stress is experienced, for the damping mechanism ($\propto e^{-kt}$) and return or recovery mechanism ($\propto \sin wt$) bring the system quickly and smoothly to R_0. A trained athlete can tolerate large S and yet show both smaller R and smaller times of return to R_0 than an untrained person. Training can therefore be thought of as increasing k, increasing w. Similarly, a diabetic can be thought of as having a decreased k and w in comparison to a normal subject for S measured as glucose dose, and R measured as blood glucose level. Stress therefore, may be thought of as that effect which occurs when a stimulus (or stimuli) is applied at such magnitude that the damping and recovery mechanisms are saturated. In other words, a complete R and S function would contain terms which will predict that, for values of $S \geqslant S_{max}$, R will not return quickly to R_0. The value of S_{max} can be determined experimentally by plotting R curves for different S. As S_{max} is approached, the resultant curves depart more and more from simple $\sin wt/e^{kt}$ curves. Since for stimuli less than S_{max}, the stress stimuli, R varies with $\sin wt/e^{kt}$, R falls in time, but for $S > S_{max}$ the effect continues; the system is stressed.

Stress then, it is suggested here, occurs when the damping and recovery mechanisms as measured by values of w and k are saturated. People with low w and k values are easily stressed; their homeostatic systems revert too slowly to equilibrium. Also, the implication here is that since w and k are finite numbers for living tissues (otherwise they would show no response at all) every person can be stressed if S is sufficiently large.

I would wish also to emphasize that several stimuli can have measurable effect on one variable of response. Thus heart rates can be varied by physical activity, by temperature changes, and by emotional stimuli. When this happens S is a function of several different stimuli, S_1, S_2 etc., each contributing to R, which may therefore reach its saturation level with relatively lower values of the stimuli independently applied. Furthermore, if R persists, as it does when the system is stressed, this value of R becomes the new quasi-equilibrium state, and further stimuli are therefore acting upon a system already removed from healthy equilibrium. This implies considerable fragility in response to new stimuli.

The merits of assuming that some form of $R \propto \sin wt/e^{kt}$ exists for all R and S relationships, I would like to suggest, lie in the observed fact that small stimuli do disproportionately affect people when they are stressed; this general phenomenon is explicable in terms in this mathematical model. It is a model to which also incorporates the fact that in terms of human health and function variables are not independent. The saturation of human recovery and damping mechanism can occur via many different paths. I have tried to show, with a minimum of formal mathematics, that since different stimuli can have effects on the same measurable system in the human body, (and we know many of these systems are interdependent) the concept of a stress capacity is one that is capable of quantitative description. For many psychological and biochemical systems in the human body the limits are known. As Bannister points out, what is stress for one person is exhilaration for another,[6] the stress of athletic sport is sinusoidal, while the stresses described by Selye[7] are often so severe that the sine function no longer operates which is in itself, I suggest, a function of unbearable stress of an over-stimulated system. The notion of discharge of tension, or stress, finds some mathematical description in the sine function. The sine function no longer operates, or is saturated, when a person is stressed. Experimental calculation of w and k in defined conditions would be a first step toward a quantitative theory of stress.

Exercises

1. Can you detect a sinusoidal effect in your measurements in Chapter 6?
2. Can you suggest a better mathematical relationship between R and S than the one briefly described here? For example

$$R \propto \frac{S.\sin wt}{e^{kt}}$$

but we require $R = R_0$, when $S = S_0$, so,

$$R = R_0 + \frac{(S - S_0) \sin wt}{e^{kt}}$$

which fulfills this requirement. Other requirements are that at some value $S > S_{max}$, R is no longer a function of S, so that the function is discontinuous at $S \geqslant S_{max}$. Can you find such a function?

Here we have considered one R and one S, but often we want to know about the effects of several stimuli at the same time. René Thom, the French mathematician, has developed a three-dimensional analysis of discontinuity called the Catastrophe Theory, which is generally applicable to R and S studies.[8]

Methods 39–40. Rhythms and the human body

The most obvious cycle in our species is birth, growth to maturity, and eventually death. Obvious, yes, but not fully understood in molecular or physiological terms. Another common cycle is the menstrual cycle, which occurs in healthy women regularly, with onset to cessation taking about a lunar month. The study of such periods is called chronobiology, and may well prove to be a very important aspect of our biology.

We know that potassium is excreted more rapidly during mid-day than during sleep. Such rhythms are termed *circadian* (about a day) because it takes them about 24 hours to exhibit a complete wave or trough (see Figure 7.3). Experiments with animals show they also have rhythms, as we would expect. Groups of mice injected with a fixed dose of endoxin, a poison, showed variable death rates according to the time of injection. The study of drug treatment and circadian rhythms shows us that, for best results, specific times in the cycle have to be chosen; jet lag fatigue is a real phenomenon, in that if you pass into a new time zone (towards or against the Sun), your circadian rhythms will obviously be out of phase with the new time zone. Although we still do not know how the mechanisms work, we have established that the endocrine system, by rhythmically altering

Figure 7.3 Simple characteristic of a periodic process. In practice the cosine (or sine) is established by measuring the parameter of interest against time, and then superimposing the degree scale.

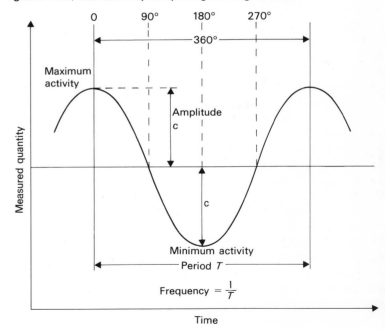

flow rates of its hormones, influences our metabolism. Thus, as Professor Halberg dramatically said (see references), transiently every normal individual has rhythmic fluctuations of adrenal activity, approximating at its peak to Cushing's disease, and twelve hours later low enough to approximate to Addison's disease.

Method 39. Identification of peak and trough values

(a) Temperature and pulse rate

You can test simply for rhythms by monitoring mouth temperature every half hour and recording the reading (to the nearest 0.1 °C) against time, but you need to keep the ambient (room) temperature fixed. At the same time, you can monitor pulse rates, which usually show a peak, with temperature, at about mid-day, but clearly you must keep the subject at the same activity, since increased muscular work causes both pulse rate and temperature.

(b) Grip and reflex

Many people feel 'brightest' in the afternoon, while physical co-ordination and strength are at minimum values at 3–4 a.m. You can identify peak and trough values of strength by plotting performance for a chosen time interval of grip strength (say 1 every hour) (see Method 35) and 'brightness' by reflex time (see Method 41).

You may discover that some of your subjects are better at night than during the day. We differ very much in our body clocks.

Method 40. To test for rhythmicity

The characteristics of a rhythmic process are shown in Figure 7.3. Usually a computer analysis of data is necessary,[9] but simple tests of periodic change can be made, as shown below. To define all the parameters exactly, you have to monitor a process as continuously as possible for several days, or even weeks.

Simple experiments to investigate chrono-rhythms

1. Measure height every hour over the longest possible period.
2. Measure height every day for someone of age 20+ years, taking care to weigh at the same time every day, with empty bladder, after bowel movement, and nude (or with exactly the same underclothes). Record weight as accurately as you can. There is some evidence that a person who maintains his or her weight actually moves a little above, then passes below, then up over again, their 'average' weight. Careful experiments of this kind are still seldom done.

These studies are still in their infancy, and as Dr Halberg commented to me:

'It would be good to see somebody become interested in actual research on this topic (effects of meal-timing) in your country'.

References
1. Adolph, E. F. in *Science and Medicine of Exercise and Sports*, edited by Johnson, W. R. (London: Harper and Row, 1960), p. 69.
2. Hallock, P. and Benson, I. C. *J. clin. Invest.*, 1937, **16**, 597.
3. Durnin, J. V. G. A. and Passmore, R., *Energy, Work and Leisure*, (London: Heinemann, 1967) p. 41.
4. Grow, M. C. and Armstrong, H. G., *Fit to Fly*, (Appleton Century Crofts, 1941) p. 199.
5. Lippold, O., *J. Physiol.*, 1970, **206**, 359.
6. Bannister, R., *Practitioner*, 1954, **172**, 63.
7. Selye, H., *J. clin. Endocrinol.*, 1946, **6**, 117.
8. A simple introduction can be found in *New Scientist*, 1975, 20 November, p. 447.
9. An excellent set of self-monitoring techniques is given by Halberg, Franz et al., Autorythmometry, *The Physiology Teacher*, 1972, vol. 1 (4), p. 298. Advanced work can be studied in current copies of *International Journal of Chronobiology*, John Wiley and Sons, London and New York.

8 Perception

Introduction to Methods 41–43

In order to perceive we require sensory organs. Our eyes, for example, are sensitive to visible light signals, and we interpret these signals as vision, but they are not sensitive to longer electromagnetic rays. In order to act on our perception we require the interpretation of the signal to result in muscular response. A reflex is strictly speaking the automatic response to a signal; for example if bright light is shone at the eye, the pupil contracts. A reflex of the type described in Method 41 is a learned response, which, if practised, becomes automatic.

Perception is, of course, a vast subject, but the approach to its study is simply shown in the methods in this chapter, which have been developed to excite an interest in studying people around you, using next to no apparatus, yet with quantifiable results.

Method 41. Measuring speed of reaction

The ability to react to a signal is of prime importance to us, since on getting information from our eyes and ears we can then judge our best reaction in physical movement to meet the situation. It would be impossible for a tennis player, for instance, to play well unless he or she could relate their strokes in time and space to the oncoming ball. Some studies have shown that ability to coordinate muscle movements in throwing and striking are very well correlated with human intelligence; in other words our brain's development during our evolution was associated with what is termed spatial awareness.

We tried to find a simple reaction test, and used a metre ruler in the following way. Ask the person to be tested to sit down, stand by him with a metre ruler held lightly by finger and thumb, so that its lower end is above his outstretched hand, and when you release the ruler it

Figure 8.1 The metre rule test to measure speed of reaction.

falls between his finger and thumb. The test is for him to catch the ruler when you drop it. The faster he reacts, the smaller the reading of fall of the metre recorded. Tell him what you are going to do, and let him have one practice attempt. You will, of course, have to standardize his method of catching, or perhaps you might wish to investigate how different people catch.

The results for fifteen men and fifteen women students were averaged for two tries each and drawn on a scatter diagram (Figure 8.2). the mean value for men was 23.2 ± 6.8, for women 23.5 ± 5.5. There is no significant difference in these results. Calculate the mean for each group on your diagram and mark in with a line.

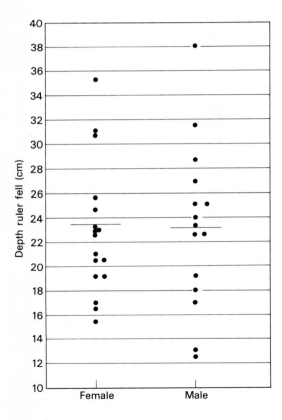

Figure 8.2 Scatter diagram of depth of fall of ruler for males and females.

Method 42. Touch, temperature and taste

(a) Touch

Areas very sensitive to touch are the mouth, finger tips and palms, and genitalia. These areas of dense distributions of receptors sensitive to pressure reflect their functions. The lips have sexual functions, while their sensitivity allows us to identify food textures. Our hands give us much information about surface and consistency of materials we touch. The genitalia when touched initiate through the messages sent to the brain the intense physiological and emotional changes for sexual intercourse.

Socially we have taboos which also reflect touch-sensitive areas. You may extend your hand to a stranger, but to touch him or her on the mouth, breast, or genitalia, is, in our culture, interpreted as a minor assault. Parents touch their very young children everywhere, but the frequency falls as the child grows. Mothers and fathers have different mutual areas of trust in their children, and this varies with the sex of the child. You may identify intimacy between people by observing where they touch each other in public. Girls holding hands attract very little attention, but eighteen year old youths can expect much bric-a-brac. An American Indian warrior demonstrated superiority in war by touching his adversary on the chest and face, or counting *coup*; they had special harmless sticks to do this, a ritual which cut casualities considerably.

Look at Figure 8.3. By shielding the site of the test area from the person studied he is dependent on his ability to discriminate between one or two points of contact by 'touch' alone. Since the calipers are allowed to rest under gravity on the skin a uniform pressure is obtained. We found that on the arm most people could discern two points at a caliper separation of 2 cm or more, but there was considerable individual variation. Also, even with wide separation some people could only detect one point of the caliper, suggesting that some points of the skin are not sensitive.

Investigate different parts of the body for sensitivity to touch in this way, and make a map of high and low sensitivity areas.

Human Measurement

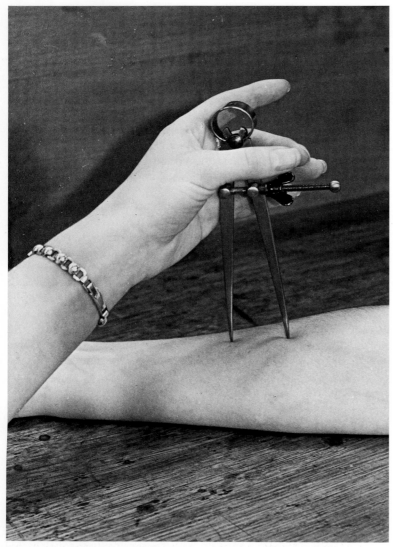

Figure 8.3 Testing the ability to discriminate between two points of contact by touch alone.

(b) Sensitivity to hot and cold

If you put your hand in water at 18 °C for three minutes, then transfer to water at 30 °C, the new immersion is distinctly identified as hot. If after 3 minutes you transfer to water at 45 °C, again a sensation of hotness is perceived. If you then transfer to 30 °C it now appears cool, or tepid. This phenomenon of subjective evaluation varying with previous experience shows how our qualitative perceptions are not trustworthy as absolute standards, but are, nevertheless, reliable in identifying differences in our experience.

(c) Perception of saltiness and sweetness

We can easily distinguish water from a solution of 0.5 g sodium chloride in 500 ml of water by taste alone, and identify the salt solution as salty. Similarly 0.5 g of sucrose in 500 ml of water is easily distinguished from water by taste alone, and the sugar solution idetifed as sweet. But how sensitive are we to these tastes? Here is a general method for quantifying ability to distinguish between water and solutions of salt or sugar (Table 8.1).

By choosing appropriate dilutions you can get very weak solutions differing very little one from the other. Is there anyone in your group who can distinguish between strength 0.05 and 0.06 in Table 8.1?

Table 8.1 Dilution values of a solution of sucrose (1 g per litre).

Solution	Sucrose solution (ml)	Water (ml)	'Strength'
1	100	0	1
2	90	10	0.9
3	50	50	0.5
4	10	90	0.1
5	5	95	0.05

Method 43. A General Method for Quantifying Scales of Performance and Perception

Scientific work is heavily reliant on observation, the perception of events, the description of matter and its interaction. From simple qualitative description comes numerical work, often by the use of measuring devices. Often we want to give a number to our description when numerical scales or measuring apparatus is either not available or inappropriate. In these cases one course of action open to us is to construct our own scales and apparatus. This is a valid approach, and has value to other people when we are able to describe our work in such a way that they can if they wish use our scales. How, for instance, would you compare people's observational powers? An interesting class experiment is to arrange for a person to simply walk into the laboratory or classroom, perform some task (e.g., tapping the radiator, sweeping up a gangway) and then depart. You can then give the class a prepared test sheet with questions of the following type:

1. What colour was the person's eyes?
2. What colour was the person's hair?
3. How long was the person here?

In this way, a score for each right answer can be given, say one for each of 10, 50, or even a hundred questions, can be given, and an 'observational' score obtained. To make assessment easier, it is common practice to give three of four choices in the answer. Thus, 'what colour was his shirt?—red, blue, green, white', would be answered by ticking the colour.

The questions of course must be unambiguous and have a single correct answer. By *giving* answers, ease of marking is assured, but you have to make sure the correct answer is objectively correct. Since the test, the timing, the person's clothes, and his actions can all be objectively described, the test can be repeated for other groups and the scores compared.

If this were done for a large number of tests, and the average score for, say, 10–11 year olds, was 60%, then for the test 60% could be given the value 100. An individual who was then given the test who scored 83% could be given the observational quotient (OQ) of:

$$OQ = \frac{83}{60} \times 100 = 138 \text{ (nearest whole number)}$$

In this manner the OQ of any child in the age group could be assessed.

Basically this is how IQ scales are constructed and measured. You will note that:

(a) The OQ does not give credit to the child who noticed things other than you asked for, and so cannot be said to represent total observational ability.
(b) The test must be prefaced by strict timetables, telling the children that a test is to be undertaken. What you say is your choice, but once decided it has always to be repeated the same way. The arbitrariness is inescapable, but it qualifies the rates of the observation you are measuring, and so again, you must conclude that you cannot say a high OQ in your test will necessarily relate to powers of observation in other fields.

This technique of producing quotients on scales is used widely in psychology; its most well known application is that of the IQ. No matter how sophisticated the method is made, it must always be remembered that the two qualifying features just discussed always apply.

Try to construct your own scales to measure performance in, for example, memory tests, or physical tests. Examine currently available IQ tests (write to the Professor of Psychology at your nearest university for further information) and observe the criteria it uses for high scoring. Can you identify any errors of omission? For example, in the OQ test I invented here, colour is asked for. A child might have excellent observation of shape, and number, but be colour-blind. If your test is biased to colour, the child will score low, and yet may have noticed all the colours he could see. You have to decide whether you are measuring colour-blindness or observation, and modify your test accordingly. IQ tests quantified by, say, a group of children picked from one social strata are inescapably qualified by that group. To use the test for a totally different group of children, differing in age, health, and so on, is immediately to bring in factors other than those you are trying to measure. This error is similar, by analogy, to comparing heights measured by the same apparatus but in one group heights were measured with shoes on, while in the other shoes were taken off. This fact should, properly, be mentioned.

Suppose you wanted to see if the OQ of girls, on your scale, was different from that of boys. You chose your age group, and gave the conditions as reproducible as possible. Suppose the mean OQ's were very different. Does this mean that the difference occurs because of sex? If the girls were all suffering from colds, and the boys were not, this single fact would invalidate your conclusions regarding sex, because for all you know the colds may be the deciding factor. You have at least two differences between the groups, sex and colds. If you dig deeper you may find even more differences, and the difference in OQ's could be, for all you know, a function of all or some of them, other than sex.

It is for this reason that chemists and physicists strive to exactly define the conditions of their experiments, so that only one factor is different when comparing a measurement. If you can be sure that the only variable between A and B is, for example, temperature, then any other difference, for example, colour, can be attributed to temperature. If then you find that as temperature varies, colour varies, the evidence is stronger that colour is a function of temperature. In human biology it is very very difficult to reach such precision.

References
Many simple perception experiments, with full statistical reasoning, are to be found in the standard text, Woodworth, R. A. and Schlosberg, H., *Experimental Psychology*, (London: Methuen, 1966).
Further simple experiments can be found in Beauchamp, K. L. *et al.* (Eds.), *Current Topics in Experimental Psychology*, (New York and London: Holt, Rinehart and Winston, 1970).

9 Some surveys and Measurements in Nutrition

Introduction to Methods 44–50

Vitamins are molecules we need, usually in milligram amounts, because we cannot make them in our body, yet they are essential participants in our vital chemical reactions in all our cells. The interesting thing about them is they are not used up, like for instance glucose which is actually broken up into smaller molecules, nor are they built into the permanent structure of cells, as are amino acids which are used to make muscle protein and tendon proteins. So why do we need to have vitamins every day in our diets if they are not used up in these obvious ways? The answer is they are lost from our systems in urine.

Vitamins are found in our blood, they get there from our intestinal tract by passing across the intestinal walls into our blood stream. Blood carries them around the body, and they pass into tissues, but they also pass through the kidneys and, as they do so, they pass across the filtration membranes into the major duct which takes them into the urinary bladder from which they are expelled. If the level of a vitamin in the blood falls because it is not being absorbed from the intestines from digested food, the vitamin continues to be lost through the kidneys. As the level falls, the vitamin is lost from our tissues too, passing from them into the blood, and thence into urine.

In this way, all our vitamins can be lost from the body, and when the levels fall specific diseases develop according to the vitamin being lost. This dynamic process is increased in the case of most vitamins by chemical reactions, distinct from the function of the vitamins, which change them chemically and make them useless as vitamins. For example, vitamin C exists as two main forms in our blood and tissues, ascorbic acid, and an oxidized dehydroascorbic acid. Both of these materials are lost in your urine, but a small amount of vitamin C is actually destroyed in our blood and tissues, eventually producing oxalic acid, which has no vitamin C activity at all, and, worse still, the degradation of vitamin C to oxalic acid is irreversible. This one-way breakdown is a general feature of vitamins; we would otherwise not be so vulnerable because we could make the vitamin back from its degraded form: we cannot do that.

One of the prices we pay for having such advanced brains and bodies to go with them is a poor synthetic ability. Bacteria do not have nervous systems but they can make most vitamins from water, carbon dioxide, ammonia and a few salts. Most mammals can make ascorbic acid easily enough from sugars, but we, with the guinea pig and some species of bat, cannot. The advantage we have is our genetic material, being free of the information the genes need to carry out these complicated processes, can contain the more advanced (in evolutionary terms) genes for an extensive nervous system.

The diseases we get after 60–90 days on a diet devoid of ascorbic acid is called scurvy. It is a terrible disease, but what is even more astonishing is that the bleeding gums, the internal bleeding, even the old scars that open up, all heal quickly when ascorbic acid is put back into the diet. Providing secondary diseases do not occur, vitamin deficiency diseases are reversible by giving the vitamin to the sufferer.

We know from very careful experiments that 10 mg of vitamin C per day, *every day*, prevents scurvy. Think of all those sailors who suffered so dreadfully with Captain Cook and Vasco da Gama, the soldiers of ancient Greece and Rome who died, and the old people who still suffer today for lack of a pinch of this white powder. Today, we have fascinating debates on vitamin C. Everyone agrees that without it you get scurvy, and everyone knows that 10 mg is the lower limit for most people to keep scurvy away, but some scientists believe, and I am one of them, that this acid protects us against the common cold. Some scientists think that it is essential for mental health too, and certainly people who, in controlled trials, do not have the vitamin appear to be more cranky than usual. Against this background of doubt riddled with certainty is the fact that none of us know exactly how ascorbic acid prevents scurvy. We know how vitamin B_2 functions in our tissues in great chemical detail but we are ignorant about ascorbic acid.

We know that the contraceptive oestrogen pill hastens the destruction of the vitamin, and that adolescent children need more than adults, but until we have identified the mechanism, in molecular terms, in which this vitamin performs, all our findings keep prodding us with the question, 'Why'? We will know one day, but I have emphasized this subject to demonstrate how very little we really know about diseases known for thousands of years, or, to put it another way, to show you what challenges there are in human measurements. Some of the experiments you can do with ascorbic acid and yourselves could in fact be producing new data.

The skin has only recently been identified as a hormone gland, producing vitamin D, which since it is carried by the blood to act on the intestines is therefore a hormone, but the absorptive properties of this large organ are still unknown. Oils have been rubbed into skin for millennia, but we know very little about the rate at which they penetrate. Here, a very simple method is described to measure rate of absorption.

Energy expenditure (E) and intake (I) must balance in the adult. If not, weight increases when $I-E$ is positive, and decreases when it is negative. For children, $I-E$ must of course be positive, since growth occurs. In this chapter, very simple yet illuminating techniques are discussed to give some idea of expenditure. The calculation of intake can easily be done if the weights and kinds of foods are known, since most foods have now been measured for their calorie content.

In practice however, the amount of detail required to get a value within $\pm 10\%$ is usually prohibitive except for full scale dietary surveys, and even then the results are usually not better than $\pm 25\%$, or worse. For this reason, I have given a meal frequency technique as a starting point for nutritional studies. Obviously, it is very approximate, but one does not have to be a nutritionist to recognize that a plateful of chips hardly constitutes a meal for a child.

Accordingly, in Method 50, some of the following need to be present for the eating to constitute a meal.
>Dairy produce (eggs, butter, milk, cheese)
>Meat or fish
>Cereal (grains, meal)
>Fresh fruit and vegetables

Again, this is only a first approximation, but it will enable those meals which are not adequate to be discounted in the counting up.

Method 44. Measurement of vitamin C in urine

In health, urine is sterile, containing no bacteria as it comes from the bladder, but it can pick up bacteria from the external genitalia. If exposed to air, urine rapidly becomes infected with a variety of bacteria which find it a hospitable place because it contains all the minerals known to be of importance in nutrition, all the amino acids, many sugars, vitamins, and urea from protein metabolism.

The measurement of materials in urine is of increasing importance (though it has always been important) since if anything is defective, or if there is change in metabolism, a change in the level and pattern of materials in urine occurs. It is used to detect early pregnancy, several inherited diseases in children and to assess nutritional levels. Very low levels of vitamins in the urine indicate a sub-optimal diet.

Ascorbic acid (vitamin C) is easily measured in urine by reacting the acid with a blue dye, usually by titration. Even as late as the early nineteen seventies, many studies were made valueless because the vitamin was not measured immediately after urine voidance. It was kept for at least a day. Unfortunately, ascorbic acid rapidly oxidizes when its solutions are exposed to air, and so these older studies were unreliable.

In 1973, doctors working in Africa found lower levels of the vitamin in women using oral contraceptives, but they did not check the urine, leaving the question open, 'Are these women excreting the acid more rapidly?' We tested this question. The experimental group consisted of 13 English women of mean age 24 years in the range

20–35. None of the subjects were obese; none were undergoing any medical treatment; none were pregnant; all were students or teachers in tertiary education. Six of the group were using prescribed oestrogen contraceptives, and all had been doing so for at least three months; and 7 had never used the contraceptives.

Urine samples (30–300 ml) were assayed withing fifteen minutes of collection. Assays were made in triplicate using standardized solutions of 2,6-dichlorophenolindophenol. The mean value (mg per 100 ml) of the urine levels of women using the oral contraceptives was found to be 1.483 ± 0.871 (12) while the mean value for women not using the contraceptives was 3.186 ± 1.339 (13). The t test ($p < 0.05$) showed there to be a significant difference between the groups. This finding suggests that for women on comparable diets the mean level of excretion of ascorbic acid is about half that of women not using oral contraceptives.[1]

The method used to estimate ascorbic acid depends on its ability, by reduction, to turn the blue dye, 2, 6-dichlorophenolindoephenol, colourless. Convenient tablets equivalent to 1 mg of ascorbic acid are obtainable from British Drug Houses Chemicals Ltd, Poole, BH12 4NN. These are crushed by pestle and mortar and dissolved in distilled water, quantitatively transferred to a flask and made up to 100 ml. This gives the so-called 1 mg % solution. Use only on day of making up and keep cool.

Collect the urine in a beaker. Fill a burette, 0–50 ml, with the urine, and titrate quickly against 20 ml aliquots of the blue dye until the dye is no longer blue. Repeat another two times. If the volume of urine is x ml, then the mg % strength of the urine is given by:

$$\text{Urine, mg/100 ascorbic acid, mg} = 20/x$$

On a good mixed diet about 3 mg are excreted per 100 ml, on a very low vitamin C diet it falls below 1 mg %, while high ascorbic acid diets give very much greater readings, e.g., 50 mg %, or more. You will note that blue dye reappears if the titrated urine + dye solutions are allowed to stand in air. For this reason the titration must be done quickly.

If your value of x is less than 5 ml, dilute your urine by a known amount before putting into the burette, since it is impossible to get accurate results for burette readings of less than 5 ml. Measure the mg % ascorbic acid in your own urine, and find the average for the class. Are the values higher before or after lunch?

Method 45. Estimate of total ascorbic acid loss in twenty-four hours in urine

You can estimate your total loss of ascorbic acid by measuring the mg % in each voidance of urine, and measuring the total volume. A group of students did this and found they were losing about 30 mg a day. How much are you losing? Tabulate as follows:

Urination episode	Vol. in ml	mg %	mg ascorbic acid = $\dfrac{\text{volume} \times \text{mg \%}}{100}$
1st			
2nd			
3rd			
5th			
5th			

To get a total, add the figures in last column. You will need to measure the first urination on waking and all those between until you go to sleep again. We have found[2] that healthy people seldom urinate more than five times a day, and for young women they excrete a total of about a litre. Very little work has been done on healthy people, so your results could be valuable.

Even the rate of formation of urine, crucial in estimating vitamin loss, has rarely been studied. Six women students monitored their urine excretion daily for six weeks. The average number of voidance episodes was 5, to the nearest whole number, and the volume was 300 ± 70 ml per episode, with a daily interval of five hours, while the fifth episode occured after nine hours, indicating a volume build up in the bladder as shown below.

Human Measurement

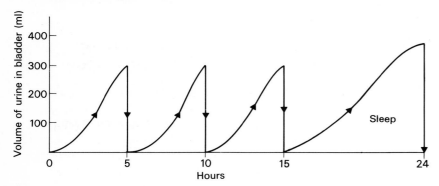

Figure 9.1 Frequency and volume of urine discharge in healthy women. Values for six women 19–20 years, height 5′ 5″ ± 1.5″, weight 122 ± 8 lb. Average volume voided during day per episode 300 ± 70 ml, after sleep 400 ± 90 ml. Intervals between episodes during the day: 5 ± 0.5 h.

Method 46. Measurement of change of excretion of vitamin C in response to change in diet

Ascorbic acid continues to be detectable in urine even a few days before scurvy begins. Using Method 44, measure your urine mg % for several urinations in one day, and then change your diet. You can increase your ascorbic acid intake by eating citrus fruits, apples, and fresh raw vegetables at every meal. You can reduce your intake by eating no fresh vegetables or fruits, and making sure your vegetables are all over-cooked. Go on the changed diet for at least three days, but not more than five. Plot your average mg % for each day against time, day one, day two, day three.

The change in vitamin C excretion usually shown by a majority of people is rising values for improved intake and falling values for reduced intake. Common findings are a reduced level of about a third the pre-diet level, and an improvement to twice the level on a raw plant produce intake supplementing a normal diet.[3]

A test group of 9 female students, aged 22–30 years, was divided into three groups: one group maintained their ordinary diet, another group embarked on a high vitamin C regimen (a), and a third group subsisted on a low vitamin C regimen (b). The diets were sustained for three days. Assays of urinary ascorbic acid were made at the beginning and end of this period. Each sample was assayed in triplicate; the variation within an assay was found to be within ±3%.

The mean value (mg per 100 ml) of the urine levels before dietary change was found to be $2.86 \pm 1.497(9)$; the control group's mean value after three days was $1.43 \pm 0.532(3)$. The high vitamin C regimen yielded a urinary assay of $5.10 \pm 0.282(3)$; and the low vitamin C regimen yielded $0.87 \pm 0.289(3)$. The t test showed ($p<0.01$) there to be no significant difference between the control group and the mean urinary level of the whole group, but significant difference between the values obtained for low vitamin C and high vitamin C regimens, and the whole group before dietary change.

The diets were as follows:

(a) Raw citrus fruits, fruit and vegetable juices, lightly cooked vegetables, with meat, and dairy produce.
(b) The low vitamin C diet contained no fresh fruit or vegetables, no fruit drinks; all vegetables were extensively overcooked.

The upsurge in vitamin C on eating plant foods poses very interesting questions. For instance, L. Pauling[4] calculated the average value of ascorbic acid in one hundred and ten plant foods, to yield 2500 kilo-calories to be 2300 mg. This is several times higher than any recognized recommended allowance. Using the same list of foods, we have calculated the amounts of several other nutrients ingested on this 'ancestral' diet, from food composition data. We found that our mean value of ascorbic acid was 3750 mg, but, more surprisingly, iron, copper, and potassium were many times more prevalent in this diet than recommended values (see Table 9.1). The conclusion is that we suffer low intakes of all the nutrients listed here, if Pauling's idea of an ancestral diet is right. We do know that women are chronically deficient[7] in folic acid, copper, and iron, and absorbtion of iron is aided by vitamin C.[8] Unfortunately, the mean weight of plant foods to give 2500 kilocalories is 9.6 kg, while the

Table 9.1 Nutrient levels associated with a plant food diet.

Nutrient	Amount[5] for 2500 kcal	Ratio of amounts to recommended or average intake[6]
Pantothenic Acid	34.25 mg	3.4
Vitamin E	53 mg	2.7
Folic Acid	1026 µg	2.6
Iron	61.4 mg	6.1
Copper	13.5 mg	6.0
Potassium	26.9 g	9.0

average western daily food intake is about 3 kg. Thus bulk poses the question, 'Is it conceivable our ancestors ate nearly 10 kg of raw plants a day?'.

The diets of aboriginal peoples (Australia) do not approach this level. We can conclude that *homo sapiens* cannot munch his way through twenty pounds weight of raw plants. However, if the calorie supplement from meat were 50% then there is a possibility of a realistic diet. As the figures in Table 9.1 show, this still implies that our diets are grossly inadequate in many vitamins and minerals.

Method 47. Can ascorbic acid in plasma be measured through the tongue?

Ascorbic acid (vitamin C) is a reducing agent; each molecule yields two hydrogen atoms for reduction. The blue dye, 2,6-dichlorophenolindophenol will easily oxidize, in a quantitative manner, aqueous ascorbic acid solutions; the product of this oxidation of ascorbic acid is dehydrascorbic acid, while the dye itself is turned into the *Leuco* or colourless form. Several workers have used this feature of the dye to show that ascorbic acid is lost through the mucous membrane of the tongue, indeed we know now that plasma ascorbic acid levels are inversely related to the time of decolourization of the blue dye 2,6-dichlorophenolindophenol upon the human tongue.[9] In other words, the faster a spot of blue dye decolourizes on the tongue, the higher the level of ascorbic acid in the subject's blood stream.

In the original papers on this method a solution of blue dye containing 0.4265 g/l (sodium salt) was used, and a table of plasma levels corresponding to the time of decolourization found, but we have found that the lingual test is really only a comparative method, and little reliance can be placed on absolute values. It can be used to identify changes in plasma-ascorbic level in individuals or between matched groups. The test is unlikely to be quantitative when comparing old people with young people, especially if the tongues are in different histological condition as so often happens with deprived senior citizens. We have also found[10] that the times for West Indian women who are ingesting the same amount of vitamin C as otherwise similar Caucasians are often in excess of 5 minutes, which would mean they were deficient in vitamin C despite very high urine values, comparable to those of the Caucasian group. The method clearly relies on identical flux properties of tongues to ascorbic acid, something which appears not to exist.

Procedure

It is essential that the mouth is well washed out with water (three times, with a minimum of 50 ml of water). Twenty microlitres of blue dye are applied to the tongue (which is thrust out), to one side of the mid-line. The time of decolourization is noted in seconds using a stop watch.

Apparatus and reagents

The pipette and tubes. 20λ micropipette tubes and pipette (by Drummond Scientific Co, Broomall, Pa., USA).

The dye. 2,6-dichlorophenolindophenol tablets (pills) are available (by BDH, Poole, UK). Each pill is equivalent to 1 mg ascorbic acid (AA). These are crushed in a mortar and pestle, and the solution made up to volume. The dye is made up in water, not acid, as BDH recommend, and the dye may also be weighed out from an AR sample of the sodium salt. A solution containing 0.4265 g per litre is equivalent to an ascorbic acid solution containing 0.258 g per litre.

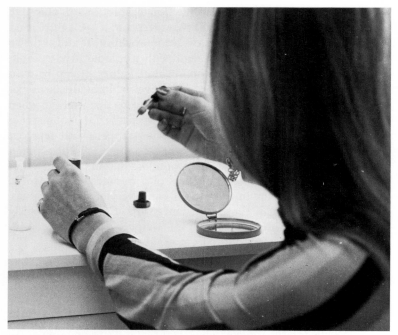 

Figure 9.2 Self administration of Lingual Test.

Calculations

The original findings were carried out using a syringe of 25 gauge, which yields approximately $20\,\lambda$. The equation below may be used for approximate estimates of plasma ascorbic acid. Although there appears a negative linear relationship between plasma AA concentration and time of decolourization for AA levels above 1.4 mg AA per 100 ml plasma (1.4 mg %) the relationship clearly departs from linearity, and, consequently, the graph shown (adapted from data of Kien and McDermott) shows this rather than a straight line more suitably for the lower plasma concentrations initially described in the literature.

When blue dye containing 0.4265 g/l is used, and the time for decolourization is less than 120 seconds, but more than (or about) 20 seconds, the equation

$$\text{plasma ascorbic acid, mg/100 ml} = -0.012\,t + 1.614$$

(where t is seconds for decolourization) may be used for monitoring the same person.

Some surveys and Measurements in Nutrition

Figure 9.3 The Lingual Test.

To see if dosage of vitamin C changes lingual times

On the day of your test do not eat breakfast or lunch; you may take coffee or tea. Record your lingual time and then ingest 1 g AA in 300 ml of water. Measure lingual time after five hours.

The lingual test is an example of a simple method which has not lived up to its promise. We have proved it works in some individuals, in the sense that their lingual time decreased with AA dosage, but there is great doubt as to the quantitative application of the method between different individuals and groups.

Method 48. Rate of absorbtion of oils and essential fatty acids through the skin

There is considerable interest in unsaturated fats (those with double bonds in their carbon chains) because some of them have been found to be essential to life just like vitamins. It is also thought that if a higher proportion of these unsaturated materials were in the diet a reduction of the formation of arterial blood clots might result. Unfortunately, one acid, *erucic*, which is concentrated in rape-seed oil, is thought to damage the heart.[11] Rape-seed oil is much cheaper than olive oil, and so has been used as an edible cooking oil.

The essential fatty acids are used in our metabolism to synthesize prostaglandins, cell regulating materials. The dietary source of essential fatty acids are olive oil, and animal foods like liver, but corn oil is used extensively in the West for frying. These acids can enter the tissues through the skin if rubbed into the skin. We did one experiment to see if the rate of absorption varied from different oils.

The skin was first washed with soap and dried. A known area was then marked on the skin (see Figure 9.4) and a measured amount of oil spread onto the skin and rubbed until a piece of filter paper put into contact with the skin showed no translucency when held up to the light. The time between this point and the addition of the oil was noted. For each oil the experiment was repeated three times. Different oils were used on different days. The results to the nearest second are shown in Table 9.2.

Table 9.2 Rates of absorption of oils into skin as a function of percentage by weight of fatty acids in oil.

Oil	Palmitoleic	Oleic	Linoleic	Time (s)	Rate × 10^5 (ml in^2 s^{-1})
Corn	8	30	53	37	10.42
Sunflower	5	30	60	30	12.73
Almond	—	67	22	210	1.82
Olive	6	80	10	130	2.94

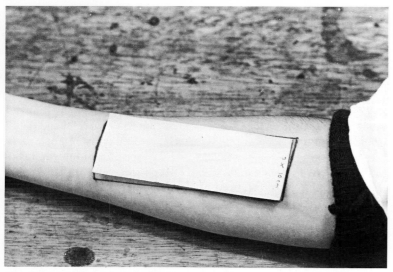 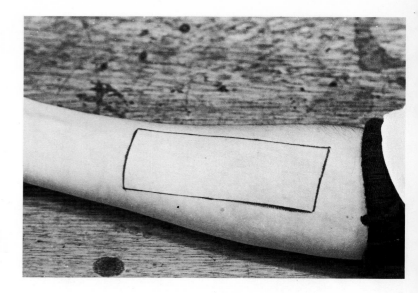

Figure 9.4 Method of marking area on skin.

Observe how the acid with two double bonds, linoleic, appears to be absorbed more quickly than the single double bond acid, oleic.

Oils are often used for sprained joints and muscles. Since linoleic is an essential acid so we must have it in our diet, it would appear that massage with such oils may have a nutritive value. (See appendix for correlation of rate and percent of specified acid.)

Method 49. Activity analysis and energy expenditure

As living beings, we still have to obey the first law of thermodynamics, which states that energy cannot be made or lost. This means that our food input must balance the work we do, the change in our weight, and the heat we lose. By measuring the rate at which oxygen is used during activity we can calculate the energy expended during the activity. This has been done for many tasks and recreations.

For ease of reference, nutritionists and physiologists use standards referred to a 25 year-old man of 65 kg, and a 25 year-old woman of 55 kg, both living at 10 °C.[12] Weight is important, because a bigger person consumes more energy than a smaller person doing the same task. Age is important because our ticking over rate, the Basic Metabolic Rate, is age variant. The temperature of the surroundings affects the rate of heat loss from the body, being higher the further it is below blood heat, and lower at higher temperatures.

Obviously, the accurate assessment of energy expenditure is very difficult indeed, but for a first approximation average rates of expenditure can be used. The monitoring of activity at first sight is a simple technique. One merely records what is being done, when it is being done, but we have to use approximations in simple survey work, otherwise the amount of data becomes difficult to handle. If we want to find the energy expended without actually measuring it at the time, the grouping together of various activities with the same energy level is a useful approximation.

Activity analysis has revealed that obese people, children as well as adults, move less often than lean people. This reduction of energy output must mean an increase in weight, if the input (food) is otherwise comparable with lean people.

A reference woman uses up 2300 kcals per day, and a reference man uses 3200 kcals. As a rough approximation, an 80 kg man will use up:

$$\frac{80}{65} \times 3200 = 3939 \text{ kcals}$$

If activity is monitored every minute results of total energy expenditure obtained by multiplying energy expenditure × time for each activity gives an agreement between 10–15% of direct calorimetric methods.

It is impossible to list here energy cost of all human activities, even where they are known, but the list below can be used, with some imagination to construct energy costs over 24 hours, using the relation:

Energy cost of activity = time in minutes for that activity × energy cost in kcals/min for that activity

and adding up for the 24 hours so that the 1440 minutes are accounted for.

Table 9.3 Energy cost in kcals/min for various activities.

Activity	Reference man	Reference woman
Walking	5.3 (6 km/h)	3.6 (5 km/h)
Working activities, standing	2.5	1.83
Washing/dressing	3.0	2.5
Sitting activities	1.54	1.41
Table tennis, cricket	2.5–5.0	2.0–4.0
Hockey, tennis	5.0–7.5	4.0–6.0
Boxing	7.5	—
Squash	7.5	6.0
Laboratory work	2.3	—
Cooking	2.5–3.7	2.0–2.9
Hanging out washing	2.5–3.7	2.0–2.9
Bedmaking	3.8–4.9	3.0–3.9
Shopping with heavy load	5.0	4.0
Typing 40 wpm, mechanical	—	1.7
electrical	—	1.5
Office work, (sitting)	1.6	—
Sleeping	1.1	0.9

Compiled from the numerous tables and references in *Energy, Work and Leisure* by Durnin, J. V. G. A. and Passmore, R., (London: Heinemann, 1967).

Method 50. Monitoring meal frequency

The human body functions best with regular meals rather than the same amount of food eaten irregularly or in fewer meals. Obesity is often associated with missed breakfast, a desultory lunch, but with heavy meals in the evening.

A simple technique for identifying eating patterns is to monitor people with the following table:

Day	Breakfast	Lunch	Tea	Dinner/Supper	Total
Monday					
Tuesday					
Wednesday					
Thursday					
Friday					
Saturday					
Sunday					

Each space is filled in with a 1 for meal taken, and 0 for no meal. The total number of meals is then found for that day. The total number for a meal can be found by summing the day totals.

Determine the number of meals people have in selected groups by
(a) asking them to fill in the diary each day as the week progresses;
(b) interviewing them each day about the previous day's meals.

These two methods will not, usually, give identical results, because people like to appear 'normal' in their answers, and because memory is not to be relied on.

Having obtained your results, try to correlate them with some other important variable. For example, is a low score correlated with smaller stature? This may be important in primary school children. Other features to be investigated would be general attentiveness and alertness; children who are sent to school without breakfast, who then eat just chips for lunch, and who do not get a meal when they go home cannot grow adequately, nor can their overall physiology be as good as it could be.

References
1. Harris, A. B., Hartley, Josephine, and Moore, Angela, Oral contraceptives and reduced urinary ascorbic acid, *Lancet*, 1973, ii, 201–2.
2. Harris, A. B., Pillay, Mary, and Moore, Angela, *Chem.-Biol. Interactions*, 1976, **14**, 371–4.
3. Harris, A. B., and Ajose, D., Rapid assay of urinary vitamin C, *Lancet*, 1973, i, 671–2.
4. Pauling, L., 1972. Diet detailed in Ballantine Book, p. 70.
5. From averaged values of calculations for foods. Pauling, Compositions from McCance, R. A., and Widdowson, E. M., 1960. MRC Special Report No. 297, HMSO London, by Donaldson, M., and Trippitt, D.
6. Recommended levels or average levels of intake averaged from Medical Research Council Special Report. Composition of Food, No. 297, HMSO 1960. Recommended Dietary Intakes of Nutrients for UK (DHSS), HMSO 1973. Davidson, S., Passmore, R., and Brock, J. F., *Human Nutrition and Dietetics*. (London: Churchill-Livingstone, 1973). *Recommended Dietary Allowances*, USA, NRC: NAS 1968. *Handbook on Human Nutritional Requirements*, WHO Monograph Series No. 61.
7. Davidson, S., Passmore, R., and Brock, J. F., *Human Nutrition and Dietetics*, 5th edn. (London: Churchill-Livingstone, 1973) pp. 422–34.
8. Brook, M., in *Nutritional Deficiencies in Modern Society*, edited by Howard, A. N. and McLean Baord, I. (London: Newman Books, 1973) pp. 45–54.
9. Kien, L. J., McDermott, *Med. J. Aust.* 1972, ii, p. 420.
10. Harris, A. B., Cooper, H., and Gardener, J., *Lancet*, 1976, i, 585.
11. Editorial, *Lancet*, 1974, 7 December, 1359.
12. Food and Agricultural Organization of the United Nations, 1957. *Calorie Requirements*, Nutritional Studies No. 15, Rome.

General references
Linus Pauling has some novel ideas on vitamins (see his *Vitamin C and the Common Cold* (San Francisco: Freeman, 1970)) while our ideas on vitamins are changing rapidly, as explored in *Orthomolecular Medicine* (edited by D. Hawkins and L. Pauling, (San Francisco: Freeman, 1973)) which is, in parts, a readable book for the non-specialist. Exhaustive studies are given in *The Vitamins*, edited by N. H. Sebrell and R. S. Harris (London: Academic Press, 1967).

Part III

Treatment of Results

10 Statistical Methods

Introduction to Methods 51–53

When you obtain results for a small group of people, properly speaking the results apply only to them and special methods have to be used to assess the relevance, or significance, of these results to other groups.

If you take a sample of the people in your college or school with no special selection procedure, it is possible to make estimates of values for the whole population of the school. When you wish to apply your results to the human race as a whole special procedures are required, involving the size of the sample, and the size of the population. These are very involved and advanced statistical techniques, and are not considered here. Only methods of treating sample data are described.

Actually the majority of research findings are related to samples, so you will not be diverging from established scientific practice. The really important point is to described your methods so clearly that other workers can reproduce your work, and directly compare their findings. If several workers find similar results produced from several sample studies, the likelihood is very strong indeed that these results are not arising purely by chance.

Luckily we have simple methods to estimate the likelihood that your results are the result of Chance, and they are described below.

Before working through these methods study Figure 10.1, which shows the most common form of the pattern of results in biological studies, the so called normal distribution. Of course not all variables are normally distributed. The number of legs we have as a species is a fixed value, but heights, weights, performance, all these vary normally about a mid, or average, point called the mean.

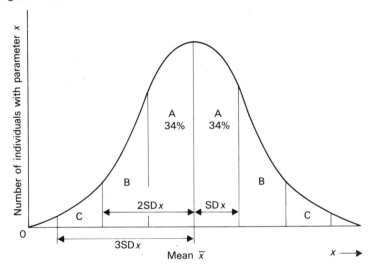

Figure 10.1 The normal distribution of x, and its standard deviation SDx. The normal curve is exactly symmetrical about $\bar{x}$. Approximately 34% of the readings are in A to the right, $\bar{x}$ + SDx *and another 34%* in A to the left, $\bar{x}$ – SDx. In other words the range $\bar{x}$ ± SDx accounts for about two thirds of the readings. The B segments account for another 13% each, and the C segments 3%.

Strictly speaking the methods described below should only ever be used for normal distributions, but it is an interesting observation that the bulk of scientific papers published hardly ever establish normal distribution of their results, yet nevertheless use these methods.

Method 51. How to establish means and standard deviations of samples rapidly

When you have collected your results it is perfectly permissible to calculate your mean and standard deviation (SD), by assuming normal distribution, unless
(a) your results are numerous and show a skew, or
(b) the distribution is known to be skewed.

In small-number determinations you are 'allowed' the benefit of the doubt. However, clearly, the more values you have, the more reliable your results, although numbers for numbers' sake is no virtue, since after a certain calculable level the labour is gratuitous. Nevertheless it is always wise when considering a variable to observe whether it is at least approximately normal. A very simple method is shown below, and as I describe it, I had no idea of the result.

The ankle circumference, above the ankle bone, of twenty women was measured using a fibre-glass non-stretch tape, and entered to the nearest millimetre. On receiving the figures I inspected them for highest (24.5 cm) and lowest measurement (20.1 cm). I then chose 0.9 cm as my frequency interval, and entered 1 against this cell everytime I found a reading in the range. (This way all the data can be checked through step by step.)

Interval (cm)	*Frequency*	
1. 20.0–20.9	1 1 1 1	4
2. 21 –21.9	1 1 1 1 1 1 1 1	8
3. 22 –22.9	1 1 1 1 1 1	6
4. 23 –24.9	1 1	2
	Total	20

Fourteen (approx. $\frac{2}{3}$ or 66%) of the readings are in cells 2 and 3. The mean will be close to 22 cm, and $\pm$ 1 cm takes in 14 readings. It follows that the mean circumference is 22 ± 1.0 cm approximately.

This sample of twenty women happened to be a class of students. As a sample there is a skew towards the smaller ankle rather than the larger; it may very well be that ankle sizes are skewed in the population in this way, but we could only estimate its probability, and rather vaguely at that, from this sample. The proper course of action would be to calculate the accurate mean, and the accurate SD from the total set of 20 readings, and present the results for that group of women only. If one wished to ascertain a value for women in the UK we would have to take a very much larger sample, in tens of thousands preferably, and using precise techniques establish whether the distribution was skew or non skew, then apply sample number correction techniques to the values obtained. Nevertheless, when working with a small group it is perfectly permissable to investigate correlation within the group, as for instance, ankle with wrist sizes. These are the dangers and advantages of working with small groups.

Essentially the normal distribution means that at any distance (or interval) from the mean, there are an equal number of examples having a value higher and lower by a specified amount than the average. The average height of young men in the USA army is 5′ 9″, there are therefore as many men between 5′ 8″ and 5′ 9″ as between 5′ 9″ and 5′ 10″ because the distribution is normal. Your results can simply be tested for normality by entering them with a cell and frequency table as I have described and finding what range takes in $\frac{2}{3}$rds of the result. When the results are numerous and show a skew, the statistical analysis is involved, and the calculation of SD inapplicable.

Method 52. Calculation of the standard deviation and *t*-test

When several hundred or more measurements are involved, the calculation of mean and standard deviation (SD) is most efficiently done on a computer. For up to 20 measurements, the method given below can be used. You can use it for any number if you have the time and patience. For medium sets there are a variety of techniques, which are all variants on the method described below, and can be found in a variety of texts.

The variable measured is x. List the frequencies as shown (I have used some figures concerning an experiment of mine, but they could equally well be any set of data produced by any experiment).

Remember to number off your headings. Class (1), $x=10$; class (2), $x=8$; class (3), $x=33$; class (4), $x=14$, class (5), $x=14$. Then square each figure to give x^2.

Table 10.1 A frequency table.

Class	x	x^2
1	10	100
2	8	64
3	33	1089
4	14	196
5	14	196
$r=5$	$\Sigma x = 79$	$\Sigma x^2 = 1645$

We now have to find the sum of x, referred to as Σx; do this by summing the x column.

$$\Sigma x = 79, \text{ called sigma } x$$

We also need the sum of the squares, so add up the figures in the x^2 column; this is referred to as Σx^2.

$$\Sigma x^2 = 1645$$

The mean is:

$$\text{mean of } x\ (\bar{x}) = \frac{\text{sigma } x}{\text{number of } x\text{'s}}$$

$$\bar{x} = \frac{\Sigma x}{n}$$

But n is given immediately from your column, i.e., $n=5$

$$\therefore \text{mean}, \bar{x} = \frac{\Sigma x}{n} = \frac{79}{5}$$

$$\text{i.e.} = 15.8 \text{ (exactly)}$$

We now require an entity d^2 which is merely a step on the way: it is given by (i) finding the square of Σx, that is $(\Sigma x)^2$.

(ii) dividing this by n, to give $(\Sigma x)^2/n$

(In our example this gives $(\Sigma x)^2 = 79 \times 79$ and $n=5$, so we get $(79 \times 79)/5 = 1248.2000$.)

(iii) taking this last figure from Σx^2 i.e. $1645 - 1248.2$, gives 396.8. This value is

$$\Sigma d^2 = \Sigma x^2 - \frac{(\Sigma x)^2}{n}$$

Divide Σd^2 by n, to give $(\Sigma d^2)/n$

e.g. $396.8/5 = 79.36$ exactly.

The SD of our set of figures (*sample*) is simply the square root of $(\Sigma d^2)/n$, i.e. $\sqrt{79.36} = 8.9084$ (to 4 decimal places).
The result is expressed as

$$x = \text{Mean} \pm \text{SD}$$

Experimental $= 15.8000 \pm 8.9084$ (to 4 decimal places).

In my experiment the result would be finally expressed as 15.8 ± 8.9 since measurements were read to the nearest whole number.

You can use this method for any set of numbers to give a mean and SD. Properly speaking it should only be used for figures of a so-called normal distribution, but providing there is no clear indication to the contrary, the calculation is an excellent means of indicating the value and spread of your results. In particular, it gives a value and limit within which about 66 per cent of your results fall. Usually it is best to use five figure logs, or a calculator. Slide rules in general cannot be used adequately for SD determinations. You also usually need five figure square root tables or a pocket calculator, too. The calculation of mean and SD is perhaps the most useful and fundamental of all statistical methods.

Mean and SD instruction sheet

(i) Number your x's, gives n;
(ii) add your x's, gives Σx;

(iii) square your x's, and add, gives Σx^2;
(iv) find $\Sigma x/n$, gives mean $\bar{x}$;
(v) find $(\Sigma x)^2$ and divide by n, gives $(\Sigma x)^2/n$;
(vi) find $\Sigma x^2 - (\Sigma x)^2/n$, gives Σd^2;
(vii) find $\Sigma d^2/n$, and take square root gives SD;
(viii) express your result $\bar{x} \pm$ SD.

The significance of the mean and standard deviation

The t-test is a simple application of statistical theory to smallish scale research of biological variables, and serves to put results into a shape which is readily distinguishable by other workers because of its standardized form.

Below I give a worked example of the t-test, whose value is to give a numerical estimate of the significance of results. To do this we calculate the value t, and by use of tables we can find out to what extent our results differ from pure chance. This estimate is universally given the symbol p. If the p value from the table is 0.01, it means that our results are 99 to 1 odds on to differ from those we would have got simply by making up a set of numbers (or to put it another way, by dipping into a bag of random numbers). If p is 0.1 it means our results are odds on by 9 to 1 to differ from a collection of random numbers.

Suppose we have two results, one for group A and the other for group B; our results are given for parameter x (height, incidence of disease, etc.) and so we express them as:

$$x_A = \bar{x}_A \pm SD_A \qquad x_B = \bar{x}_B \pm SD_B$$

where $\bar{x}$ is the mean, and SD the standard deviation.

We want to know if x_A differs significantly from x_B. If $x_A = x_B$ then there is no significant difference, but even when $x_A \neq x_B$, if the interval ends $\bar{x}_A + SD_A$ and $\bar{x}_B - SD_B$ overlap, then there is diminishing chance that there is a significant difference in the results. If however, these outermost limits do not overlap the difference is very significant indeed. It is useful to draw on a graph your results, as shown, where you can judge the amount of overlap visually before doing your sums (see Figure 10.2).

We must now use the formula, which I have purposely cast in terms of SD's.

$$t = \frac{\text{difference of means}}{\sqrt{[SD_A^2/n_A + SD_B^2/n_B]}}$$

where n_A and n_B refer to the number of observations or readings in each group. In this form you can use t for your results in their final form and they refer to your samples, in other words to the sets of figures you had.

The working that follows is to four decimal places for convenience, but five places are safer. Below is an application of the t-test.

The mean value (mg per 100 ml) of the urine levels of ascorbic acid of women using the oral contraceptives was found to be $X_A = 1.483 \pm 0.871$ while the mean value for women not using the contraceptives was $X_B = 3.186 \pm 1.339$ where $n_A = 6, n_B = 7$.

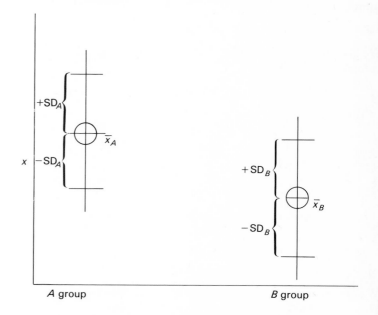

Figure 10.2 Generalized visual inspection diagram.

Difference of means = 3.186 − 1.483 = 1.703

$$\frac{SD_A^2}{n_A} = \frac{0.871^2}{6} = .1264 \quad (a)$$

$$\frac{SD_B^2}{n_B} = \frac{1.339^2}{7} = .2561 \quad (b)$$

$$(a) + (b) = .3825$$

Sq. Root

$$(a) + (b) = .6185$$

$$t = \frac{1.703}{.6185}$$

$$= 2.7536$$

There are tables which relate this value, t, to the number of degrees of freedom and the value p, explained earlier (see Tables 10.2 and 10.3). In general, the number of degrees of freedom is one less for each set of measurements. n_A is one set, n_B is another set, so the total number of degrees of freedom is $(n_A + n_B - 2)$. You now consult such tables and find p. At $n_A + n_B - 2(= 11)$, the nearest p was 0.02 for $t = 2.72$, and one is justified in giving results as significant to 1 in 50, but it is safer to use $t = 2.20$, giving $p < .05$.

It may therefore be concluded that at the $p < 0.05$ level there was a significant difference in the values of ascorbic acid found in the two samples of people used. A result which has $p < 0.001$ is generally regarded as certain, while $p \leq 0.05$ is regarded as the upper limit of significance, and in biological work is often taken to denote results significantly different from chance.

Exercises

(a) Calculate the significance of the results in Table 4.1.
(b) Given mean foot area of six women to be 129.96666 ± 30.86406 cm² and that of six men to be 179.28333 ± 17.22212 cm², find whether the difference is significant. Observe that these values are calculated to physically meaningless decimal places in order to do the statistics.

Table 10.2 Distribution of t.

Degrees of freedom	Probability, p				
	0.1	0.05	0.02	0.01	0.001
1	6.31	12.71	31.82	63.66	636.62
2	2.92	4.20	6.97	9.93	31.60
3	2.35	3.18	4.54	5.84	12.92
4	2.13	2.78	3.75	4.60	8.61
5	2.02	2.57	3.37	4.03	6.87
6	1.94	2.45	3.14	3.71	5.96
7	1.89	2.37	3.00	3.50	5.41
8	1.86	2.31	2.90	3.36	5.04
9	1.83	2.26	2.82	3.25	4.78
10	1.81	2.23	2.76	3.17	4.59
11	1.80	2.20	2.72	3.11	4.44
12	1.78	2.18	2.68	3.06	4.32
13	1.77	2.16	2.65	3.01	4.22
14	1.76	2.14	2.62	2.98	4.14
15	1.75	2.13	2.60	2.95	4.07
16	1.75	2.12	2.58	2.92	4.02
17	1.74	2.11	2.57	2.90	3.97
18	1.73	2.10	2.55	2.88	3.92
19	1.73	2.09	2.54	2.86	3.88
20	1.72	2.09	2.53	2.85	3.85
21	1.72	2.08	2.52	2.83	3.82
22	1.72	2.07	2.51	2.82	3.79
23	1.71	2.07	2.50	2.81	3.77
24	1.71	2.06	2.49	2.80	3.75
25	1.71	2.06	2.49	2.79	3.73
26	1.71	2.06	2.48	2.78	3.71
27	1.70	2.05	2.47	2.77	3.69
28	1.70	2.05	2.47	2.76	3.67
29	1.70	2.05	2.46	2.76	3.66
30	1.70	2.04	2.46	2.75	3.65
40	1.68	2.02	2.42	2.70	3.55
60	1.67	2.00	2.39	2.66	3.46
120	1.66	1.98	2.36	2.62	3.37
∞	1.65	1.96	2.33	2.58	3.29

Table 10.2 is abridged from Table III of Fisher and Yates: *Statistical Tables for Biological, Agricultural and Medical Research*, Oliver & Boyd Ltd, Edinburgh, by permission of the authors and publishers.

Table 10.3 The correlation coefficient, *r*.

Degrees of freedom	Probability, p				
	0.1	0.05	0.02	0.01	0.001
1	0.988	0.997	1.000	1.000	1.000
2	0.900	0.950	0.980	0.990	0.999
3	0.805	0.878	0.934	0.959	0.991
4	0.729	0.811	0.882	0.917	0.974
5	0.669	0.755	0.833	0.875	0.951
6	0.622	0.707	0.789	0.834	0.925
7	0.582	0.666	0.750	0.798	0.898
8	0.549	0.632	0.716	0.765	0.872
9	0.521	0.602	0.685	0.735	0.847
10	0.497	0.576	0.658	0.708	0.823
11	0.476	0.553	0.634	0.684	0.801
12	0.458	0.532	0.612	0.661	0.780
13	0.441	0.514	0.592	0.641	0.760
14	0.426	0.497	0.574	0.623	0.742
15	0.412	0.482	0.558	0.606	0.725
16	0.400	0.468	0.543	0.590	0.708
17	0.389	0.456	0.529	0.575	0.693
18	0.378	0.444	0.516	0.561	0.679
19	0.369	0.433	0.503	0.549	0.665
20	0.360	0.423	0.429	0.537	0.652
25	0.323	0.381	0.443	0.487	0.597
30	0.296	0.349	0.409	0.449	0.554
35	0.275	0.325	0.381	0.418	0.519
40	0.257	0.304	0.358	0.393	0.490
45	0.243	0.288	0.338	0.372	0.465
50	0.231	0.273	0.322	0.354	0.443
60	0.211	0.250	0.295	0.325	0.408
70	0.195	0.232	0.274	0.302	0.380
80	0.183	0.217	0.257	0.283	0.257
90	0.173	0.205	0.242	0.267	0.338
100	0.164	0.195	0.230	0.254	0.321

Table 10.3 is abridged from Table VII of Fisher and Yates: *Statistical Tables for Biological, Agricultural and Medical Research,* Oliver & Boyd Ltd, Edinburgh, by permission of the authors and publishers.

Method 53. Correlation

When we determine values of x with their corresponding y's we often wish to decide if x and y are in any way related. The soundest procedure is simply to draw a graph of x versus y. If the graph looks as if there is a straight line relationship between x and y, then the next step is to determine the quantitative relationship between your readings by a process termed *calculation of coefficient of correlation*, which I describe in detail here. If however, your graph shows no obvious relationship between x and y then the chances are that you have a random relationship between x and y, which put more precisely means 'given any value of y you still may not predict its x value with certainty'. The proper course of action is then to accept this as an experimental fact, a not unimportant finding; or you may insist that there is, lurking behind the chaos, a pattern.

However, in the simpler cases the relationships stand out rather clearly, and, let me add here, the more care taken in organizing your experiment to keep, as near as possible, all interferring factors invariant the fewer measurements you need and the stronger your findings. If your x–y curve appears clearly as a curve other than a straight line, then you may not sensibly calculate the coefficient of correlation. Obviously, if the curve is clearly shown, there is a quantitative predictive relationship, but you may not use the following method. I give below a detailed description of finding the correlation between the width of the human pelvis and ankle circumference.

Ten women, all 19 years of age with triceps skinfold between 1.5 and 1.8 cm, all Caucasian, and none more than ± 1.5 SD of height outside mean female height, were measured using a non-stretch fibre tape. The question was 'is there a significant correlation between hip width and ankle circumference?'. After obtaining the results we proceeded as follows.

(i) Tabulate x–y pairs in two columns.

Item	x	y
1	21.0	23.0
2	21.0	26.0
3	21.0	21.0
4	22.0	26.5
5	21.0	21.0
6	21.3	23.6
7	23.0	27.2
8	22.5	25.3
9	20.1	23.3
10	21.8	23.8
$n=10$	$\Sigma x = 214.7$	$\Sigma y = 240.6$

(ii) Find the sum of x, i.e. Σx, and the sum of y, Σy. Find $\bar{x}$ and $\bar{y}$.

$$\bar{x} = \frac{\Sigma x}{n} = \frac{214.7}{10} = 21.47$$

$$\bar{y} = \frac{\Sigma y}{n} = \frac{240.6}{10} = 24.06$$

(iii) Tabulate x^2, y^2 and xy (the product of x and y) from the data in step (i).

Item	x^2	y^2	xy
1	441.0000	529.0000	483.0000
2	441.0000	676.0000	546.0000
.	.	.	.
.	.	.	.
.	.	.	.
10	475.2400	566.4400	518.8400
Totals:	$\Sigma x^2 = 4616.1900$,	$\Sigma y^2 = 5830.8200$,	$\Sigma xy = 5176.6900$

These three steps give all the data we require to find the correlation coefficient.

Calculate the entity $\Sigma d^2(x)$, which is defined as

$$\Sigma d^2(x) = \Sigma x^2 - \frac{(\Sigma x)^2}{n}$$

which, by substituting values we find is

$$4616.19 - \frac{(214.7)^2}{10} = 6.5810$$

Calculate the corresponding $\Sigma d^2(y)$, defined as

$$\Sigma d^2(y) = \Sigma y^2 - \frac{(\Sigma y)^2}{n}$$

$$= 5830.82 - \frac{(240.6)^2}{10} = 41.984$$

Calculate the entity $\Sigma d(x)d(y)$ defined as

$$\Sigma d(x)d(y) = \Sigma xy - \frac{\Sigma x \Sigma y}{n}$$

which in our example has the value

$$5176.69 - \frac{214.7 \times 240.6}{10} = 11.008$$

We now observe that the correlation coefficient is given by

$$r = \frac{\Sigma d(x)d(y)}{\sqrt{\Sigma d^2(x)\Sigma d^2(y)}}$$

$$\therefore r = \frac{11.008}{\sqrt{6.581 \times 41.984}} = .66224$$

There is therefore a positive correlation of x with y, so that as x increased so does y. The value $\Sigma d(x)d(y)$ defines the sign. Using tables of p versus r we find that for $p = 0.05$, $r = 0.602$ for 10 measurements (nine degrees of freedom). (If you have one measurement you have no degrees of freedom since it is fixed; if you have two measurements, when one is fixed, the other can still vary, so, in general, the number of degrees of freedom must always be one less than the set of measurements.) We therefore conclude that the correlation is a true one, and we can be sure that the results could only occur by chance once in 20 times. Biologically this result shows that we can expect larger hip widths in women with larger ankles, and vice versa.

Now that we have the value of the coefficient of correlation we must find the quantitative expression relating x and y. This is called the regression equation, and may be very easily calculated as shown below.

Find the means $(\bar{x}, \bar{y})$ of x and y, defined as

$$\bar{x} = \frac{\Sigma x}{n} \ ; \ \bar{y} \ \frac{\Sigma y}{n}$$

which in our example gave

$$\bar{x} = \frac{214.7}{10} = 21.47, \text{ and } \bar{y} \ \frac{240.6}{10} = 24.06;$$

and calculate the factor f, defined as

$$f = \frac{\Sigma d(x)d(y)}{\Sigma d^2(x)} = \frac{11.008}{6.5810}$$

The regression equation is simply given by:
$$y = \bar{y} + f(x - \bar{x})$$
which in our example has results:
$$y = 24.06 + 1.67269 (x - 21.47)$$
i.e. $\quad\quad\quad y = 1.67269x - 11.85265$
or Hip width $= 1.673 \times$ (ankle circumference) $- 11.853$.

It is natural to ask 'does this straight line equation allow us to predict hip widths from ankle circumference?'. Let us see. In this example, once we had calculated the equation we stopped the first three available female students and obtained real hip and ankle data; we then calculated from our equation the expected results, which are all shown on the Table 10.4.

It will be seen that the results are very good indeed. It is absurd to expect a 'better' result, because we have already proved by finding the value r that the value of hip width and ankle circumference is not exactly fixed. Our regression equation therefore gives us *likely* values. It is perfectly permissible to use regression equations in this way, but they may not be extrapolated beyond the range of measurements taken to construct the equations.

References
There are so many excellent extensive and comprehensive standard texts in statistics it is invidious to mention one. Many professional scientists refer to M. J. Moroney, *Facts from Figures* (London: Pelican, 1951), reprinted consistently since then into the seventies, as an easy-to-follow introduction to the subject. Another much used text is Bradford Hill, *Principles of Medical Statistics* (London: Lancet Publications, 1971). For a series of brilliant papers in this relatively new field of mathematics, see Karl Pearson's *Early Statistical Papers* (Cambridge: CUP, 1971).

Table 10.4 Table to show difference between actual and expected ankle measurements.

Hip width (cm) y	Ankle circumference (cm) x	Predicted hip width	% Predicted / Actual	% Error
25.5	22.0	24.95	97.85	−2.145
21.0	20.5	22.44	106.87	+6.80
23.6	21.3	23.78	100.8	+ .80

11 Correlations in Classifications of Physique

Introduction

All methods of classifying physique have their limitations. If too many measurements are needed, the method becomes too cumbersome for any but small scale studies, while if few quantitative values are used, the method may become too subjective to gain confidence.

In this chapter, the background to physique classification is discussed, its importance suggested, the theoretical grounds for various systems evaluated, and correlations between methods described in the text are calculated.

Strengths and weaknesses in methods of classifying physiques

The suspicion that some physique types may be vulnerable to certain diseases has a long and confused literature.[1] Early classifications[2,3] failed, either because of their rigidity or their subjectivity. Constitutional studies were in 1954 meagre,[1] and continue to be so despite obvious interest in the subject psychiatrically,[4] in medical sport analysis,[5] in human biology,[6] in obesity,[7] and in the study of Olympic athletes.[8] The importance of the subject has been strongly stressed by Sheldon,[3] who recognised that 'in ordinary clinical work, we collect enough fragments of knowledge about people to make up a highly respectable body of science, but we are foiled by the difficulty of keeping records of it systematically.' Just as clearly, biochemical assays, physiological readings, and drug evaluation, all lose some significance when not related to the specific human body from which they emerge. Unfortunately, anatomy has suffered a decline,[9] while available techniques to classify physiques are faulty or too cumbersome. Sheldon described[3,10,11] how a physique may be considered as made up of varying proportions of three extreme types of physique; the endomorph, the mesomorph, and the ectomorph, but never succeeded in escaping the charge of subjectivity. We can now see how some of these difficulties can be overcome.

Are balanced physiques more prevalent, and what does this imply?

When the ectomorphy of a 4-4-4 is decreased by one unit to 4-4-3, with the implied increase of bone and joint size, the incidence rises from 4% to 6%, perhaps implying that the 4-4-4 is more fragile than 4-4-3, or perhaps less fertile. A decrease of fat and abdominal size from 4-4-4 to 3-4-4 yields a more common physique (5.7% as against 4%). The most common of all male physiques, according to Sheldon's last analysis on 46 000 men, is the 4-4-3 (6%), the next are 3-4-4 (5.7%), 3-5-3 (5.6%), and 3-4-3 (5%), while the 4-4-4 is sixth in order of frequency. The male population is clustered round 4-4-4 within ±1 of each component. The separation (in 3D space) of any physique x-y-z from 4-4-4 is given by:

$$\sqrt{(x-4)^2 + (y-4)^2 + (z-4)^2}$$

or approximately by

$$\sqrt{[(\tfrac{\Sigma x}{n})-4]^2 + [(\tfrac{\Sigma y}{n})-4]^2 + [(\tfrac{\Sigma z}{n})-4]^2}$$

where summation is taken over groups of very similar physiques. This gives a measure of the mean distance from the balanced 4-4-4 physique of x-y-z physique. The effect of changes in x, y and z on the prevalence of a physique is shown below in Table 11.1. Thus the further displaced a physique is from the balanced physiques, the less common it is. For this reason we chose as a standard the 4-4-4 physique, and conceived of the other physiques being generated from it (see Figure 11.1) rather than using Sheldon's concept of mixing three extremely rare or atypical physiques.

Human Measurement

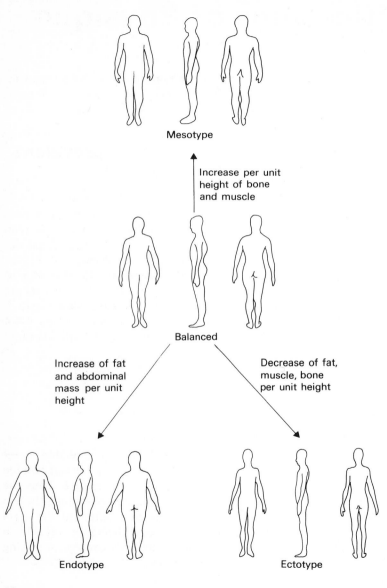

Figure 11.1 This analysis gives four basic physiques for constitutional research. A further three can be theoretically generated by considering mixtures of the three extreme physiques to give endo-meso, ecto-meso, and ecto-endo-types. Here we can think of some extreme physiques being derived from a statistically average physique, and so we do not need to consider any arbitrary types of physique as Sheldon did as a basis for classification, but use an entirely quantitative and objective method.

Table 11.1 Effect of changes in x, endomorphy, mesomorphy, and ectomorphy on the prevalence of a physique.

Incidence % between	Distance from 4-4-4 in 3D space
4–5	1.29
2–3	1.97
0.5–1	2.21
0.2–0.3	3.59
0.02–0.06	3.95

Percentage figures corrected to first significant figure.

Method 54. The case for seven body-types

Sheldon was aware that a method which divided the male population into 76 different physiques might very well be too cumbersome, and proposed a simpler division into nineteen different types; but this is still rather large, especially when some types are very rare indeed; accordingly I give here ten basic physiques, seven common and three very rare, covering the whole population each characterized by a name which distinguishes it.

The rationale behind Table 11.2 is (a) collection of similar physiques using Sheldon's original method, (b) calculation of mean somatotype for each group, and (c) the assignment of the body-type number for each category from (b). Seven common physique types, amounting to 97.8% of the male population, emerged. Three others

Table 11.2 Conversion table for components and the seven categories of body-type.

Body-types		Means (to the nearest 0.5)			Percentage incidence
Name	Components	Endomorphy	Mesomorphy	Ectomorphy	
Endo-types	622, 522, 533, 721, 731, 532, 523, 631, 621, 632, 712, 613, 612, 623, 642	6±1	2±1	2±1	5
Meso-endo-types	451, 542, 541, 442, 461, 452, 551	4.5±0.5	5±1	1.5±0.5	12
Meso-types	262, 252, 353, 271, 371, 361, 261, 362, 352, 172, 163, 162, 263, 253	2±1	6±1	2±1	18
Ecto-types	226, 225, 217, 216, 316, 326, 127, 126, 136, 236, 235, 335	2±1	2±1	6±1	10
Meso-ecto-types	153, 244, 145, 245, 254	1.5±0.5	4.5±0.5	4.5±0.5	10
Endo-ecto-types	524, 514, 515, 424, 325, 415, 425	4±1	1.5±0.5	4.5±0.5	3
Balanced (equi-types)	433, 343, 453, 354, 334, 435, 345, 444, 434, 344, 534, 443, 543	4±1	4±1	4±1	42

The component classifications are grouped together according to the proportions of the three components. On calculating the means of these components, the ranges applicable to each category was found. When the 7-category division of human physique is required, the full component classification, second column, can be used to identify the category by inspection. The incidences were calculated from Sheldon's data on men.

are endomorph (7-1-1), mesomorph (1-7-1), and ectomorph (1-1-7) which have incidences of 0.01%, 0.03%, and 0.02%, respectively. If the body-type assignations and their members are used as a basis of constitutional research then, apart from the three rare extremes, the seven body-types cover 100% of the population. One of the advantages of using the body-typing scheme lies in its possible use as a very rapid means of categorizing physiques, for whereas it is unlikely anyone could ever carry nineteen categories in their heads, seven is quite feasible.

Table 11.3 Incidence of body-types in 175 college women.

Body-type	% incidence
Balanced	24
Endo	14
Ecto	24
Meso	10
Endo-meso	14
Ecto-meso	8
Endo-ecto	6

Data calculated from findings of Bullen and Hardy, showed the following somatotypes not identified in Sheldon's original list of 76 somatotypes; meso-ecto: 1-4-6, 2-3-4, 2-4-3; endo: 5-3-1; endo-ecto: 3-2-4, 4-2-3; balanced: 3-3-3; endo-meso: 3-4-2, 4-3-2. A 2-4-3 type is shown in Figure 3.10.

Table 11.4 Anthroscopic incidence of body-types in 310 college women (Harris).

Body-type	% incidence	Mean body-type
Balanced	23	4-4-3.5
Endo	13	5.5-3-1.5
Ecto	8	1.5-2-6
Meso	11	3-5.5-1.5
Endo-meso	13	5-5-1
Ecto-meso	28	2-4.5-4.5
Endo-ecto	4	4.5-1-5

Percentages have been rounded up or down to nearest whole number.

If the figures of Table 11.4 are compared with the incidence data of Bullen and Hardy in Table 11.3, agreement is found, except for the types shown in Table 11.5.

It appears that the incidences have been reversed in Table 11.5. The explanation lies in Bullen and Hardy using the male standards to classify female physiques.

Table 11.5 Differences between Harris' incidences of women's body-types and those of Bullen and Hardy.

	% incidence	
Type	Harris	Bullen and Hardy
Ecto	8	24
Meso-ecto	28	8
Total %		
ecto+meso-ecto	36	32

Method 55. Statistical difficulties

Sheldon used five different areas of the body in his somatotyping and meticulously gives these five areas a somatotyping rating. Thus, although the final somatotype rating is presented as a numbered code, for example, the common 4-4-3 (No. 745 in Atlas), it is actually $(3.9 \pm .54)$-$(3.9 \pm .22)$-$(3.2 \pm .44)$, if the five regions are averaged. This yields highest range 4.5-4-3.5 and lowest range 3.5-3.5-3, to the nearest half unit, or 4-4-4 and 3-4-3 if we take the result to the nearest whole number.

This qualification is generally applicable to all somatotypes except the extremely rare physique which shows identical rating throughout the five regions. If equal weight is given to the regions, as Sheldon did, then the final result without the standard deviations is arbitrary.

Body-typing and somatotyping

We have carried out several hundred body-typings on young women. The body-typing for three balanced, non-obese young women is shown in Table 11.6. The last example is noteworthy for the very low values of standard deviations, indicating homogeneity of physique; the first example, (a), has a very commonly encountered amount of dysplasia ('mixing'), as shown by SD's, but it is not extreme.

Table 11.7 shows how the Sheldon somatotype is really much less precise than his final published value suggests (see last column).

Table 11.6 Body-typing for three balanced, non-obese young women.

(a) 3.25 ± 1.41	4.29 ± 1.29	3.30 ± 1.27	Balanced (3-4-3)
(b) 4.48 ± 1.06	4.30 ± 0.96	4.35 ± 0.98	Balanced (4-4-4)
(c) 3.93 ± 0.68	2.75 ± 0.43	2.81 ± 0.81	Balanced (4-3-3)

Table 11.7 Relationship between body-typing and somatotyping.

Sheldon	Means of Sheldon somatotype	Body-type assessment
2-4-4[1]	(2.0±.2)-(3.5±.4)-4.0±.0)	(3.4±1.2)-(3.6±1.3)-(4.7±.8)
3-6-2[2]	(3.4±.2)-(6.1±.2)-(2.0±.0)	(2.5±.5)-(5.2±.8)-(2.3±.7)
6-3-1[3]	(5.8±.7)-(3.4±.2)-(1.6±.4)	(5.1±1.3)-(3.9±1.2)-(3.0±1.2)

Examples drawn from Atlas, 1: no. 235, 2: no. 603, 3: no. 1097. We have calculated the mean assessment (2nd column) of Sheldon's own assessment; the third column shows how the more descriptive body-type assessment contains the more rigid somatotype.

Objectivity and subjectivity

When a method depends on measurements, it is commonly referred to as objective, whereas if it depends on the recognition of types, or forms or estimates without measurements, of relative proportion, it is commonly referred to as subjective.

What would your response be if asked, 'What are your statistical criteria for identifying people as either men or women?' Would you suppose the question absurd, since you tell men and women by anatomical inspection, or to put it more familiarly, by looking at them? Do you suppose that measurements, any set of conceivable measurements, to *identify* men and to *identify* women, could be justified in comparison to simply using a visual approach? Evidently not, since if statistical criteria were needed for categorization we would be forced to carry with us a measuring rod to tell the difference between a mouse, a man, and a mongoose. Once shown a picture, or better still, an example of a mouse, you will not confuse it with a wolf.

Pictures can be extremely useful standards. To recognize mesomorphy is simply done once you have seen photographs of mesomorphic people, or once you have had living mesomorphs pointed out to you. Having established visual standards, verbal descriptions are capable of precision too. A male *Homo sapiens* is distinguishable from the female in having testes and penis; the female has breasts and vagina. The features are defining criteria; measurement is not necessary.

Why there can be a quantitative classification system

Just as the surface of an egg has the approximate equation

$$\frac{x^2}{a^2} + \frac{y^2}{b^2} + \frac{z^2}{c^2} = 1$$

where x, y and z are co-ordinates in space and a, b and c are constants, so the human body must also be describable in three dimensions. This leads us to expect a purely quantitative method, but the difficulty lies in making it simple enough to use. We have already seen that endomorphy, mesomorphy and ectomorphy, when expressed as measurements, overlap. Consequently, we have to depart from these ideas if we want a purely quantitative technique.

The ponderal index

The ponderal index (height/$\sqrt[3]{\text{weight}}$) was used as a guide in assessing the somatotype of a physique by Sheldon (1940). It has been remarked by Harrison *et al.*, in 1964, that Sheldon's quantitative method does not work since it relies on width/height ratios. Below, the inherent contradiction in using the ponderal index as a criterion in somatotyping at the same time as using height/width ratios is analysed.

Variation of weight with height for fixed ponderal index

If the ponderal index is to remain constant for a given somatotype of weight, w, and height, h, then

$\quad w = kh^3 \quad$ where k is the cube of the reciprocal of the ponderal index (1)

$\therefore dw = 3kh^2 dh \quad$ (2)

$\therefore \dfrac{dw}{w} = \dfrac{3\,dh}{h} \quad$ (3)

$\quad dw = \dfrac{3}{h} w.dh \quad$ (4)

or, approximately, $\Delta w \simeq \dfrac{3w\Delta h}{h} \quad$ (5)

i.e., percentage increase in weight is approximately equal to 3 times percentage increase in height.

Variation of weight with variation of height with constant width/height ratios

For simplicity consider the dimension changes in the leg. The widths of the ankle both frontally and laterally must remain in constant proportion to the height,

$\quad$ Width of ankle frontally $= k_1 h = b_1 \quad$ (6)
$\quad$ Width of ankle laterally $= k_2 h = b_2 \quad$ (7)
$\therefore db_1 = k_1 dh \quad$ (8)
$\quad db_2 = k_2 dh \quad$ (9)

For simplicity, take the case $b_1 = b_2$.
For a thin segment of this cylinder, of length l, the weight for jth segment is given by

$\quad w_j = \pi \dfrac{b_1^2 K}{4} \quad$ where K is the density $\quad$ (10)

We can consider the body to be made up of such cylinders,

$\therefore W = \sum w_j = \sum \dfrac{l\pi b^2 k}{4} \quad$ (11)

$\quad = \sum \dfrac{l\pi h^2 K}{4k} \quad$ (12)

where l is any length to any degree of accuracy (e.g. in cylinders of length $l = 1$ cm) i.e. l is a constant; so is π, and K is approximately 1 g/cm³, while k depends on the somatotype and region of body;

$\therefore w = \dfrac{l\pi K}{4} \cdot \sum \dfrac{h^2}{k} \quad$ (13)

$\therefore dw = \dfrac{2l\pi K}{4} \sum \dfrac{h}{k} dh \quad$ (14)

$\therefore dw = \dfrac{2dh \Sigma h/k}{\Sigma h^2/k} \quad$ (15)

$\therefore \dfrac{dw}{w} = \dfrac{2dh}{h} \quad$ (16)

i.e., percentage increase in weight = twice percentage increase in height.

Since equations (5) and (16) are not identical, it follows that the two criteria used by Sheldon are contradictory. Since Sheldon used width ratios, it follows that his ponderal indexes must vary for the same Sheldon somatotype, or vice versa.

Results

If a somatotype at height h has weight w then its ponderal index is

$\quad \text{PI} = \dfrac{h}{\sqrt[3]{w}} \quad$ (17)

If height increases to $h + \Delta h$, then weight, if width ratios are constant, changes to

$\quad w + \dfrac{2\Delta h}{h} \cdot w = w\left(1 + \dfrac{2\Delta h}{h}\right) \quad$ (18)

$$\therefore \text{PI} = \frac{h + \Delta h}{\sqrt[3]{w(1 + 2\Delta h/h)}} \quad (19)$$

$$\frac{\text{PI}(h + \Delta h)}{\text{PI}(h)} = \frac{h + \Delta h}{\sqrt[3]{w(1 + \frac{2\Delta h}{h})}} \cdot \frac{\sqrt[3]{w}}{h} \quad (20)$$

$$= \frac{(1 + \Delta h/h)}{\sqrt[3]{(1 + 2\Delta h/h)}} \quad (21)$$

For example, put $\frac{\Delta h}{h} = \frac{5}{170}$, a 2.9% increase in height, then $\frac{\Delta h}{h} = \frac{5}{170} = \frac{2.94}{100} = .0294$

Now we must substitute $\Delta h/h$ for % increase divided by 100 in our general equation.

Ratio of PI by change x% in height is given by

$$\text{Ratio} = \frac{1 + x/100}{\sqrt[3]{(1 + 2x/100)}} \quad (22)$$

Table 11.8 Changes in ponderal index on height increase.

% height increase	Ratio
1	1.0033
2	1.0067
5	1.0171
10	1.0351

Table 11.8 shows the variation of these ratios with height increase.

From Sheldon's Atlas, somatotype 1-5-4 has

$$h = 70'', W = 138 \text{ lb}, \text{PI} = 13.55.$$

If we increase height to 75″, i.e. $\frac{\Delta h}{h} = \frac{5}{70} = .0714$

new PI $= 13.55 \times 1.0247$
$= 13.8856$

weight at 75″ $= \frac{h^3}{\text{PI}^3} = 157.58 \text{ lb}$

The weight listed by Sheldon for this somatotype (Atlas, p. 51) is 171 lb. If PI is used as a measure of somatotype, as Sheldon does in his Atlas, it follows that the width criterion has been jettisoned. On the other hand, if the width/anthroposcopic method is used, then much lighter weights are to be expected in the same but taller somatotype than listed by Sheldon. Indeed, Sheldon's listing of 171 lb for 75″ height for somatotype 1-5-4 at age eighteen has PI = 13.51.

From Sheldon's Atlas, somatotype 3-7-2 has
1. $h = 70$ $w = 185$ PI $= 12.29$
2. $h = 75$ $w = 222$ PI $= 12.39$

while if ratios are to remain equal the PI at 75″ is
$12.29 \times 1.0247 = 12.59$ giving a weight of 211.4 lb.

Sheldon made his taller examples of the same somatotype *heavier than they should be*.

If circumference/height ratio measurements were made the criteria of somatotypes the contradiction would be removed, since cross-sectional area is proportional to circumference squared, and, therefore

$$\text{weight} \propto \text{sum of terms of (height} \times \text{circumferences}^2) \quad (23)$$

If height/circumference values are used as a criteria of somatotyping then for every body circumference, c, there is a constant q, such that
$$h = qc, \quad (24)$$
for a given somatotype.

Equations (23) and (24) imply weight $\propto h^3$, in other words $h/\sqrt[3]{w}$ would be constant. This is why I used the circumference method for classifying physiques, as used in the text.

Method 56. Correlation of body-index and body-type

1. Women. From the data in Table 11.9, the parameters in Table 11.10 are calculated.

The following correlations were then calculated.
Correlation coefficient of body-index (BI) with: endomorphy = .9571; mesomorphy = .7615; ectomorphy = −.8548. The regression equations are:

(I) BI and Endomorphy
Body Index = $0.5560 \times$ (Endomorphy) $+ 0.3560$

(II) BI and Mesomorphy
Body Index = $0.7571 \times$ (Mesomorphy) $+ 1.0999$

(III) BI and Ectomorphy
Body Index = $-0.4250 \times$ (Ectomorphy) $+ 3.7375$

Table 11.9 Mean values of morphology of women of different body-types.

Mean body type	4-4-4	1.5-3-6	3-5-2.5	2-4-5	5-5-2
Number of subjects	14	6	5	13	10
Mean age, years	20	23	28	24	22
Mean height (cm)	165	165	169	167	167
Mean PI	42.69	44.87	42.52	42.96	41.53
Mean circumferences (cm)					
Bust	90	84	91	86	98
Waist	69	65	69	69	78
Hips	95	90	98	94	102
Wrist	15.6	14.5	16.0	15.1	16.6
Ankle	21.4	21.0	22.3	21.4	22.3

Table 11.10 Correlation of body-index and body-type for women.

Body-type	4-4-4	1.5-3-6	3-5-2.5	2-4-5	5-5-2
Body-index	2.2	1.2	2.0	1.6	3.4

Correlation of body-index with ponderal index

Correlation coefficient of Body Index with PI = -0.8786
PI = -1.2881 (BI) $+ 45.5933$
or
BI = -0.5993 PI $+ 27.8013$

References
1. Tanner, J. M., *Lancet*, 1956 (Sept. 29), **ii**, 635.
2. Kretschmer, E., *Physique and Character*, (London: Kegan Paul, Trench and Trubner, 1925).
3. Sheldon, W. H., Stevens, S. S., and Tucker, W. B., *The Varieties of Human Physique*. (New York: Harper and Row, 1940).
4. Kellet, J. M., *Lancet*, 1973, **i**, 860.
5. Khosla, J., *Lancet*, 1974, **i**, 30.
6. Harrison, S. A., Weiner, J. S., Tanner, J. M., and Barnicot, N. A., *Human Biology*, (London: OUP, 1964).
7. Seltzer, C. C., and Mayer, J., *J. A. Am. Nutr. Assoc.* 1969, **55**(5), 454.
8. Tanner, J. M., *The Physique of the Olympic Athlete*, (London: Allen and Unwin, 1964).
9. Sinclair, D., *Lancet*, 1975, **i**, 875.
10. Sheldon, W. H. J., *Hist. Behavioural Sci.*, 1971, **7**, 115.
11. Sheldon, W. H., Dupertuis, C. W., and McDermott, E., *Atlas of Men*, (New York: Harper and Row, 1954).
12. Bullen, A. K., and Hardy, H. L., *Am. J. Phys. Anthrop.*, 1946, **4**(1), 37.
13. Brock, J. F., *Lancet*, 1972, April, 701.

Table 11.11 Correlation of body-index and body-type components in men and women.

Body-index correlated with:	Endomorphy	Mesomorphy	Ectomorphy
Correlation coefficient	.9393	.8356	−.9477

Data from Table 3.15

Appendix
Calculations on Data in the Exercises

Here you will find fuller analysis of some of the results discussed in the text under the exercise headings. You can observe how the statistical methods discussed previously are applied.

Method 13. Correlation of armspan and height in men and women

Data used shown in Method 13.

For 16 women, 18–24

Armspan (cm) = 1.0319 × (height, cm) + 4.3891
Correlation coefficient 0.7994, $p < 0.001$

For men

Armspan (cm) = 0.8863 × (height, cm) + 23.1272
Correlation coefficient 0.7898, $p < 0.001$

The probability, p, that these results were pure chance is less than one in a thousand.

Method 19. Analysis of physiques analysed in Table 3.2

x, y, z stand for the components score, and f the number of times scored.

	Endomorphy		Mesomorphy		Ectomorphy	
	x	f	y	f	z	f
	1	13	1	0	1	2
	2	13	2	0	2	6
	3	3	3	2	3	7
	4	3	4	30	4	18
	5	0	5	1	5	1
	6	0	6	0	6	1
	7	0	7	0	7	0
Mean	1.88		3.97		3.37	
Standard deviation	±.94		±.31		±.18	
Body-type	2		4		3	

Method 21. Body-indexing (data from Table 3.14)

Subject	1	2	3	4	5	6	7
Mean body index	1.56	1.33	2.00	1.89	2.11	2.67	2.56
Standard deviation	.53	.50	.71	.78	.93	1.12	.88
Body index	2	1	2	2	2	3	3

Method 35. Correlation of strength and grip

If N is the number of times a 14 pound weight is lifted in 30 seconds, and G is the measured score on a grip metre in scale units (1 scale unit ≃ 1.4 lb)

Mean value $N = 11.4000 \pm 2.6749$
$G = 19.6000 \pm 2.8751$
Correlation coeff., r, $= 0.9332$
∴ equation is $G = 1.0031\,N + 8.1645$

When $N = 10$, $G = 18.20$ (to second decimal place).

This allows the regression line to be drawn (see Method 35), when $n = 10$, $G = 18.20$ (to second decimal place).

For $r = .9332$ at 9 degrees of freedom the probability that the results are pure chance is less than one in a thousand, or to put it another way, the correlation is very significant indeed, so that we are justified in using grip strength as a prediction value, that is, the stronger the grip the stronger a person is. Observe that in calculating the regression equation the physically meaningless $N = 11.4$ is used; when substituting N in the equation, clearly a whole number is used.

Method 48. Rate of oil absorption correlated with percentage of specified fatty acid in oil

Data from Table 9.2.

Fatty acid	Correlation coefficient	p
Palmitoleic	.0375	not significant
Oleic	−.9423	<.02
Linoleic	+.9569	<.02

Anthropometric Instruments

The majority of exercises in the text can be carried out using metric rulers, fibre glass tapes, and large set squares. A little ingenuity can produce excellent data. Instruments are expensive, but a good range is produced by:
Holtain Ltd, Crosswell, Crymmych, Pembrokeshire, U.K. (these have automatic gauges giving lengths to the nearest mm) and Siber Hegner & Co. Ltd., 14 Talstrasse, 8002 Zurich, Switzerland.
Grip meter. A useful meter (with a calibrated but non-dimension scale) is made by Rank Medical Supplies.
Calipers. Free samples can in some cases be obtained from Servier Laboratories Ltd., U.K.

Index

Acids
    ascorbic, 77–83, 92–3
    dehydroascorbic, 77, 81
    deoxyribonucleic (DNA), 36
    erucic, 83
    fatty, 83, 106
    folic, 80
    lactic, 60, 62
    linoleic, 84
    oleic, 84
    oxalic, 77
activity analysis, 84–5
adolescent growth spurt, 10
ambidexterity, 16
anatomy and mathematics, 14, 69–70, 89–96
anthropometric instruments, 106
arm span in relation to height, 14–15, 105
ascorbic acid (vitamin C), 77–83, 92–3
    in plasma, 81–3
    in urine, 79–81, 92–3
asymmetry of limbs, 15

Bannister, R., 69
basal energy expenditure, 49
Basic Metabolic Rate, 84
    and eating patterns, 86
Bernard, Claude, 67
blood glucose levels, 67–9
blood pressure, 60–1, 67–8
body-indexing, 32–5, 104–5
body-types
    art and, 37
    assessment of, 17, 31
    child, 32
    female, 30–2, 100
    percentage, 28, 99
    sexual dimorphism and, 37
    standard, 24
body-typing
    advantages of, 32
    circumference as criteria in, 28, 30–1
    correlation of body-index and, 104
    fat percentage in, 49, 58
    procedure for, 17–23, 26
    quantitative scales for, 29, 32
    rating, 23, 26
    scoring system for, 18–20, 23–6, 31, 98
    see also body-indexing, body-types
bone size in physique
    classification, 50, 52
Bullen, A. K., 100

Calculation of correlation coefficient, 94–6
Catastrophe Theory, 70
catecholamines, 60
Chibber, S. R., 15
chronobiology, 70
chrono-rhythms, 71
circadian rhythms, 70–1
circumference
    body-types and, 28
    measurements, 30–2, 40, 103
    standards for women, 30–2
    surface areas and, 54–6
    weight changes and, 57–9
compression of joints, 4
contraceptive pills, effect on vitamins, 78–9
copper and diet, 80
correlation, calculation of coefficient, 94–6
correlation of body-index and body-type, 104
correlations in physique, classification, 97
curves, percentile, 11
cycles, human, 70–1
    see also chronobiology
        chrono-rhythms, circadian rhythms

Dehydroascorbic acid, 77, 81
diabetes and blood glucose levels, 67–9
diet,
    and iron, 80
    and minerals, 78, 81
    and vitamins, 78–9, 81–3
    see also nutrition measurement
dimorphism, sexual, 36–7, 40
DNA (deoxyribonucleic acid), 36
Durnin, J. V. G. A., 49, 56–7, 68
dysplasia, 31, 100

Eating patterns, 86
ectomorphs, 17–23, 28–37 *passim*, 40, 59, 97–104 *passim*
ectomorphy rating, 23
endocrine system, 70–1
endomorphs, 4, 17–23, 28–37 *passim*, 40, 97–104 *passim*
endomorphy rating, 23, 29
energy
    activity analysis and, 84–5
    expenditure, 49, 78, 84–5
    intake, 78
energy cost, 85
erucic acid, 83
exercise, effect on pulse rate, 60–5

Fat density and lean body-weight, 56–7
fat percentage, 49–50
fatty acids, 83, 106
feet, measuring area of, 45
folic acid, 80
frequency distribution in histograms, 7–8
frequency intervals, 90

Genetics
    growth rates and, 3, 12
    handedness and, 15
    synthetic ability and, 77
glucose levels in blood and urine, 67, 77
glycogen, 67
glycolysis, 62
glycolytic mechanism, 62
Gray, G., 10, 11
grip strength, measurement of, 16, 44, 63–5, 71, 105–6
growth, differential, 13
growth curves, 11
growth patterns, 9–11
growth rates
    genetics and, 3, 12
    nutrition and, 14
    sexual differences and, 11, 13, 36
    socio-economic class and, 12, 13
    see also height, weight

Halberg, Dr Franz, 71
handedness, 15–16, 44
hands, measuring area of, 44
Hardy, H. L., 100
Harrison, S. A., 101
Harvard Fitness Test (modified), 66
head, measuring area of, 56
heart recovery, 61
height
    arm span in relation to, 14–15, 105
    categorization by, 35

chrono-rhythms and, 71
differential growth of, 13
distribution of, 7–8
measurement of, 3–5, 9–11, 44
obesity and, 49–50
ponderal index and, 29–30, 101–2
prediction of, 12
sexual differences and, 44
surface area and, 54
histograms, use of, 7–8
homeostasis, 67–71 *passim*
homeostatic mechanisms, 67–9
human time cycles, 70–1
    see also chronobiology, chrono-rhythms, circadian rhythms

Iron and diet, 80

Joints, compression of, 4

Kien, L. J., 82

Lactic acid in the blood, 60, 62
lean body-weight, 56–7
limbs, asymmetry of, 15
lingual tests, 81–3
lingual times, 83
linoleic acid, 84

McDermott, E., 82
mathematics and anatomy, 14, 69–70, 89–96
Mayer, Professor J., 32
measurement of area
    feet, 45
    hands, 44
    human body, 49
    trunk, 56
mesomorphs, 17–23, 28–37 *passim*, 97–104 *passim*
mesomorphy rating, 23

metabolism, basic rate of, 84
minerals and diet, 78, 81
mongols, physique classification of, 28
morphological typing, 29, 97
    obesity and, 49
    of women, 30, 104
    see also body-types, body-typing, ectomorphs, endomorphs, mesomorphs, physique classification, physiques

Nutrition measurement, 77–81
    eating patterns and, 86

Obesity
    activity analysis and, 84
    assessment of, 59
    eating patterns and, 86
    fat percentage and, 49–50
    identifying, 49–50
    localized, 59
observational powers, *see* perception
observational quotient, 75–6
oleic acid, 84
Olympic events
    limits in performance, 61–2
    prediction of trends, 63
oral contraceptives, effect on vitamins, 78–9
oxalic acid, 77
oxygen loss during activity, 62, 84

Passmore, R., 68
Pauling, L., 80
percentile curves, 11
perception
    interpretation of signals, 72
    measurement of, 74–6
    subjective recognition by, 101
photography
    classification of physiques by, 32
physiological performance

limits, 61–2
measurement, 60–1
physique classification, 17–35, 99–103
  analysis of, 105
  artists and, 37–9
  bone-size and, 50, 52
  correlation of, 97
  photography in, 32
  sexual dimorphism and, 36–7
  silhouettes and, 52
  see also body-indexing, body-types, body-typing, obesity, physiques
physiques
  balanced, 26, 97
  female, 30–1
  self assessment of, 26
  see also body-indexing, body-types, morphological typing
plasma ascorbic acid levels, 81–3
ponderal index, 29–31, 101–4 passim
prostaglandins, 83
protein
  metabolism, 78
  muscle and tendon, 77
puberty
  growth rate in, 10, 13
  sexual development in, 12, 36
pulse rate, 60–1, 63–4, 67–8, 71

Rape-seed oil, 83
reaction, measuring speed of, 72–3
regression equations, 96, 106
rhythms, human and animal, 70–1

Scurvy, 77–8, 80
Selye, H., 69
sensitivity of touch, 74
sexual development in puberty, 12, 36
sexual dimorphism, 36–7, 40
shape, measurement of, 40
Sheldon, Dr William, 17, 23, 28–9, 97–103 passim
silhouettes, physique classification by, 52
Singh, I., 15
sinusoidal curves, 67, 69
skeletal measurement, 52
skin, absorption by, 78–84
skinfold measurement, 49–50, 54
  and lean body-weight, 56
socio-economic class
  growth rates and, 12
  puberty and, 13
somatotyping, 32, 98, 100–3
spatial awareness, 72
standard deviation
  body-typing and, 100
  calculation of, 90–1
  significance of, 92–3
statistics
  frequency curves, 90–1
  methods of achieving, 89–96
  normal distribution, 89, 90–3
stature, see height
stimulus, continuous application of, 68–70
strength
  body weight and, 63, 65
  correlation of grip and, 105–6
  definition of, 63–4
  fitness testing and, 66
  measurement of, 63–71
  pulse rate and, 63, 71
stress, types of, 69–70

Sun Alliance Ideal Weight Tables, use of, 49, 52
surface area, measurement of, 54

Tanner, J. M., 10–11
taste, comparative perception of, 74
temperature
  energy expenditure and, 84
  measurement of body rhythm by, 71
Thom, René, 70
touch, as a means of perception, 73–4
touch sensitive areas, 73
trunk, measuring area of, 56

Unsaturated fats, 83
urine
  loss of vitamins in, 77
  measurement of vitamins in, 78, 81, 92–3
  rate of formation of, 78–9

Vitamins and diet, 81–3
  effect of contraceptive pill, 78
  levels in urine, 78–9, 81

Weight
  chrono-rhythms and, 71
  differential growth of, 13
  identification of ideal, 54, 57
  lean body-, 56–7
  measurement of, 3–8
  obesity and, 49–50, 52–4
  ponderal index and, 29–30, 101–2
  sexual differences and, 44
  strength in relation to, 63, 65
  surface area and, 54